T0284862

Praise for ***Lessons from the Climate Anxiety Counseling Booth***

"Brimming with practical strategies to engage your heart, mind, and body in the work of climate justice, this book is a roll-up-your-sleeves resource for finding purpose, community, and even joy in uncertain times. Schapira's superpower is her rich experience working with activists, mental health experts, and frontline communities. Here, she distills the resulting wisdom in a guidebook that is equal parts balm and ballast for the work ahead."

—Sarah Jaquette Ray, author of *A Field Guide to Climate Anxiety: How to Keep Your Cool on a Warming Planet*

"Going beyond the lessons themselves, Schapira offers thoughtful questions and engaging practices to help us embody the personal and collective transformation needed in these climate changed times."

—LaUra Schmidt, founder of Good Grief Network and author of *How to Live in a Chaotic Climate: 10 Steps to Reconnect with Ourselves, Our Communities, and Our Planet*

"Schapira knows how fear can trap us. She feels that fear, too. In *Lessons from the Climate Anxiety Booth*, Schapira offers a path out: Here is how we can stop turning away from our beautiful planet, from our hot and unknowable future, and turn toward them instead. Here is how we can be brave."

—Eleanor Davis, author and illustrator of *The Hard Tomorrow*

LESSONS FROM THE CLIMATE ANXIETY COUNSELING BOOTH

LESSONS FROM THE CLIMATE ANXIETY COUNSELING BOOTH

HOW TO LIVE WITH CARE AND PURPOSE IN AN ENDANGERED WORLD

KATE SCHAPIRA

NEW YORK

Hachette Go, an imprint of Hachette Books
Hachette Book Group
1290 Avenue of the Americas
New York, NY 10104
HachetteGo.com
Facebook.com/HachetteGo
Instagram.com/HachetteGo

First Edition: April 2024

Published by Hachette Go, an imprint of Hachette Book Group, Inc. The Hachette Go name and logo are trademarks of the Hachette Book Group.

Library of Congress Cataloging-in-Publication Data

Name: Schapira, Kate, author.
Title: Lessons from the climate anxiety counseling booth: how to live with care and purpose in an endangered world / Kate Schapira.
Description: First edition. | New York, NY: Hachette Go, an imprint of Hachette Books, 2024. | Includes bibliographical references and index.
Identifiers: LCCN 2023041088 | ISBN 9780306831676 (hardcover) | ISBN 9780306831683 (trade paperback) | ISBN 9780306831690 (ebook)
Subjects: LCSH: Climatic changes—Psychological aspects. | Anxiety—Treatment. | Schapira, Kate.
Classification: LCC BF353.5.C55 S33 2024 | DDC 152.4/6—dc23 /eng/20231204
LC record available at https://lccn.loc.gov/2023041088

ISBNs: 978-0-306-83167-6 (hardcover), 978-0-306-83169-0 (ebook)

Printed in the United States of America

LSC-C

Printing 1, 2024

CONTENTS

INTRODUCTION

YOU ARE HERE

"What can *I* do about climate change?" she said. "I'm only one person."

We were in downtown Providence, Rhode Island, in the Narragansett homeland, across from the bus station, and not far from the river. The spring air was a little chilly still, but nice in the sun, with pigeons, sparrows, and humans walking around. This human had stopped to talk with me because I was sitting behind a sign that read CLIMATE ANXIETY COUNSELING 5¢: THE DOCTOR IS IN. She'd stopped for the same reason I'd started sitting out there, inviting conversations with strangers: because our knowledge of climate change and our worries about it were too much to hold alone.

In the days that followed, multiple people stopped to ask me similar questions and gave that same name to their anger, their fear, their despair: *I'm only one person. What can I do?* It turned out that none of us were just one person: there were a lot of us. And the truth that we're not alone gives us our best chance of not just living but living well in the hard times that climate change is bringing.

Summer after summer is the hottest on record. Crops fail. Roads melt. People's homes are flooding, burning, blowing away. That was happening in 2014 when I had that first conversation. But back then it was possible to ignore if it wasn't happening to you—and even if it was, the fact that it was a result of climate change wasn't always part of the discussion. No one I knew seemed to want to discuss it at all, and that made me feel frantic and alone. I tried to push my reactions away, disconnected and unable to think of any action that would matter. I'd pause under a too-green oak and picture a spring when its leaves would not come back. I was conflicted when my friend told me she was pregnant: I wanted to celebrate this new person but was angry and sad about how precarious climate change will cause their life to be. And I couldn't figure out how to talk about any of it. But I needed to talk about it. More than that, I needed to listen: to know what other people were thinking and feeling about climate change, and whether they felt as helpless as I did, or had answers I hadn't considered.

This book offers you what I have learned since then—through years of asking, listening, answering, and asking again—about how to treat our emotional responses to climate change not as dead ends, but as living beginnings. This book is for you if you recognize climate change as an escalating situation that won't go away, are struggling with your own responses to it, and want those responses to be more caring, purposeful, and just. It will guide you in transforming your feelings around climate change—anxiety, loneliness, anger, fear, grief, uncertainty, exhaustion, and more—into ways to connect and reasons to act.

I can put that into words now, but when I took a leap and set up the Climate Anxiety Counseling booth, I didn't have a way to talk about it. I barely had a plan. The idea came from the comic strip *Peanuts*, where Lucy offers PSYCHIATRIC HELP for a nickel. I thought that people at bus stations, parks, outdoor markets, and other public places might get the reference. Maybe a sign reading CLIMATE ANXIETY, a term that at that time I'd never heard anyone else use, would make them curious enough to stop and talk with me. But I had no idea how those conversations would go, or where they might lead.

In my first few weeks at the booth, friends, students, acquaintances, and strangers all stopped to talk: an economics major, a mother of a young son, a gigging musician, someone in recovery. They shared nightmare visions of an unlivable future climate and stresses that weighed on them in the present. They showed me guilt, optimism, despair, and only occasionally denial or dismissal. A stranger in his forties told me the days of his boatyard were numbered by sea level rise. He didn't expect any help from the state or federal government to relocate the business out of the range of storm damage—"120 employees, they're not gonna do anything for that." And a young guy with an overlay of Spanish and Rhode Island accents spoke with frustration and fear about his three-year-old son's future: "How do people not at least care to not make it worse?"

Listening to them, I began to hear how complex our reactions to climate change are, and how our efforts to protect ourselves from the worst feelings get in the way of protecting each other from the worst consequences. Because climate change is not only huge in itself but interacts with other huge social and political forces in our lives, it can seem like nothing we could do could possibly be enough to respond to all of it. Most of all, what I learned at the Climate Anxiety Counseling booth was how alone and therefore weak so many people felt in their fear and their trouble.

And yet, through its moments of openness and connection, the counseling booth also got its visitors thinking and talking about how connecting with others might help them live better and do more. Two of my students stopped by to discuss the best place for their climate activism efforts—policy work or grassroots organizing? A teenager spoke quietly about her hopes for clean wind energy in the Port of Providence, near her school. Visitors came up with skills they could share and ideas for distributing resources among more people, or told stories of caring for the ecosystem directly, like one man who carried endangered fish over a dam (in laundry baskets!) to the place where they mate. I was surprised and encouraged, again and again, by how generously people responded to the climate-related changes they felt and the ones they feared.

It was through the booth and the people who spoke to me there that I began to learn about the many paths to making change on purpose, and how to find my place within them—paths that this book will help you identify in your life and near your home. The changes we make inside ourselves are entwined with changes in our actions. For me, that meant joining the local fight against a liquefied natural gas facility and volunteering with local land stewardship organizations; it meant reaching out beyond my communities, to be guided by people meeting and making change in more ways than I could learn at home. It's also led me to become more involved in peer mental health work in my city and state. Like so much else about our climate-changing present, being with someone in their pain can be frustrating or frightening: like all care, it can also be an honor and a gift, a way of making meaning and purpose. As you identify your purpose within climate change, and locate the care you can offer and accept, your own transformation will flow outward to other people, to your communities, and to the living world you're part of: plants and animals, fungi and bacteria, the land and the water and the air.

The questions I found myself asking and answering over and over, and the wisdom and courage that people shared with me through our conversations, have taken shape in this book: I and others have led them as workshops, trainings, and rituals. My years of developing this process and practicing it with others—six years at the counseling booth, and more since then—have shown me that it works. Here are the elements this book will use to help you feel your way into climate and community emotional intelligence, and transform what you feel, with others, into connection and action.

Stories and insights from people I spoke with at the Climate Anxiety Counseling booth and from people whose work on climate, trauma, and justice has been important and effective. Real people at all stages of the process of surviving and shaping deep changes have agreed to the sharing of their stories with the readers of this book. (Where people aren't named, it's usually because they spoke to me at the booth, where I offered anonymity.)

Questions to guide reflection, alone or with others, and help you to recognize your own climate emotions and how they shape your choices; to identify your current roles in communities and systems, including ecosystems; and to imagine how you might change your actions and your roles to meet climate change and its effects—even its worst effects—with justice and compassion.

Practices and activities that will help you build and share resilience, courage, and trust. I designed or adapted these based on what people at the booth told me they were missing, lacked confidence for, or wanted to get better at. Even visualizing or planning these practices will bring you closer to surviving and compassionately shaping our future on a shared, changed planet. Doing them, especially with other people, is even better.

Witness and analysis from me: how embarking on this process has allowed my ecological grief, fear, and anger to move me toward climate and community action, informed by the recognition of humans' interdependence with the rest of the living world—including each other.

We each react to the full knowledge of climate change in our own way and at our own pace, and our experiences of climate change also differ according to the way the world has treated us. I ask you to look for yourself in the stories here, paying attention to what they call forth in your body, mind, memory, and imagination; to answer the questions, respecting that others' answers may differ from yours, and using the process as a chance to listen as well as to speak; to try out the practices, even when they make you feel strange or vulnerable, and return to the ones that offer you the greatest feelings of strength and connection. Specific, concrete guidelines for doing the questions and practices start on page xix. There's a glossary of words and terms on page 221—words that aren't common in ordinary conversation or that have particular meanings when we're using them to talk about climate change—to make it easier to proceed with a shared understanding.

The best way to travel through this book is in order and together. In order, because the progression of stories, questions, and practices is designed to unwind tightly tangled grief, frustration, exhaustion, and inertia—the state that keeps us doomscrolling or shrugging or veering away into dissociation or panic or punching down—into a followable path of courage, capability, and strength. Together, because while transformation begins with our own responses to our changing world, it doesn't and can't stop there. To create the large-scale social and structural transformations we need, our individual concerns must move us toward effective and inclusive group action—including ways of doing and being that we haven't yet imagined.

Lessons from the Climate Anxiety Counseling Booth will help you start wherever you are, with whomever you're with. You might want to form a new group with shared climate concerns as a focus and this book as a guide, or fit the questions and practices into relations and connections you've already built: your coworkers, your fellow parents, your recovery or chronic illness support group, your tenants' association or building co-op. Let yourself imagine how strange it might feel, at first, to talk about climate change in some of these contexts, with some of these people—how being alone might seem preferable. And imagine how, as you talk and listen with greater care, witness intense feeling, and prepare for relevant, possible, practical action, you'll learn more about how to work together for more good days for all of us.

As I wrote this book, I asked myself every day the questions I want you to feel strong enough to ask yourself: *Is this what I should be doing with my time, my labor, and my love? How can I change, and how should I change?* And I continued to be haunted by the questions that may, also, be haunting you: *Does any of this matter? Are we fucked, are we doomed, is the world about to end?*

Whether it is or not, it's here now, and you are here, and we are here together. If I have to intelligently and imaginatively transform my life to meet the changing world, I want company. I want what I wanted from the very beginning, when Climate Anxiety Counseling wasn't a thing and the reality of climate change felt at the same time all-encompassing and out of reach: to feel the truth that I am not alone. Individual helplessness becomes group effort, grief becomes possibility, as we let our world transform us into who we need to be, together, now.

GUIDELINES FOR THE QUESTIONS AND PRACTICES

You can do the questions and practices solo or with one or more people, and you can imagine doing them or actually do them. If you're not able or not ready to physically do the practices, you'll still get something out of imagining or planning them. Actually doing them with other people will probably feel the least comfortable, but also the most supportive in practicing change.

You can, and should, also adapt the exercises to suit your culture, context, experience, physical and emotional capability, and your group size and purpose if you're doing them in company.

SOLO

- Read over the questions and practices a couple of times before you start to do them. If you think that one of them will be incompatible with your well-being, skip it or change it.
- Have snacks and water around or eat something beforehand.

- You can answer the questions inside your head, out loud, or in writing/typing/text. Have a notebook or a recording device handy in case it helps you to write or talk your responses out, and in case you want to refer to them later.
- If you have steadying or grounding exercises that you do regularly, do one of those before you start the questions and practice, and anytime during the exercise that you need to. (There are also grounding exercises throughout the book.)
- Try to bring your attention back to the place you're in. You may go "in and out" in your attention, and that's fine—just bring it back.
- Some of these exercises ask people to be more present in their bodies. Skip these if you already know that they're not good for you. If you discover that partway through, stop and change some aspect of your surroundings or sensations (listen to a song, splash cold water on your face, move to a different room, pet or hold an animal).
- Wind down at the end of the exercise by noting some part of the experience (a feeling, fact, vision, perception) that you want to leave behind, and something you want to take with you.
- If a practice is something you already do regularly, offer it to someone else you think will like it, or invite them to do it with you.

IN A GROUP

- Your group should contain people you already have some trust and familiarity with. Some of the exercises are harder than others, different ones will be hard for different people, and not all of them will work for every group.

- The guidelines for doing the questions and practices solo are all useful here as well! Share the questions and practices a few days before you do them, to give people time to get used to them. And if everyone will enjoy it, share a meal at the end, or eat at the same time if you're doing the exercises remotely.
- You can ask one person to lead the exercise (ask the questions, describe the practice) or take turns leading—agree which it will be before you begin.
- Learn what will make the location and the exercise accessible to everyone who's gathering, and set up those conditions. Some resources for doing so are named at the end of these guidelines.
- It usually works best for everyone to answer one question before moving on to the next.
- Whatever ways you have of organizing yourselves and being responsible to each other while you're together also apply here—codes of conduct, community guidelines, etc. If you don't have such things, developing them before you begin would be a good idea, including what will happen if someone goes against them once or repeatedly.
- Take breaks. Any person can also leave temporarily or for the duration, for any reason.
- Remain in control of what you reveal, and don't push others. People's different histories may make these questions and practices difficult for them in different ways and amounts. Choosing a story to share, thinking in an unusual way, remembering, and feeling can all be stressful or painful.
- Social safety is dynamic. The things that create it within a group shift based on many factors; it's possible to feel

unsafe without being in danger, and possible for what you do to scare, anger, or silence someone without you meaning or wanting to. If you notice yourself or someone else responding as though there's a threat, pause, slow down, and see if a person or the group needs to make an alteration or a course correction.

- For practices that involve interacting with people outside the group, everyone must assess the risk as it applies to them and their willingness to take that risk, remembering that interactions and situations can change unexpectedly.

AS PART OF ANOTHER EVENT OR GATHERING

- The two sets of aforementioned guidelines are also useful here.
- If it's a short gathering or if you have other things to work on, limit yourselves to one question set or one practice.
- When you're choosing an exercise, you can choose one that feels especially relevant to your usual purpose for gathering, or one that feels like it will add to or deepen your usual ways of interacting.
- If your group already meets regularly for another reason, you can use the ways of organizing yourselves and looking out for each other that you already have. But you can also take this opportunity to add or change some of those ways, if that feels like it will create a better time for everyone involved.
- Here is some guidance on making events accessible, from Cornell University: https://accessibility.cornell.edu/event-planning/accessible-meeting-and-event-checklist.

- And from The Mighty: https://themighty.com/topic/disability/accessibility-questions-travel-food-allergies-chronic-illness-disability.
- Leah Lakshmi Piepzna-Samarasinha also has a great access rider, and some expansive thoughts on *what* everyone should be able to access, in their book *The Future Is Disabled*. COVID safety is also an access issue; masked in-person gatherings and online gatherings are two safer options to keep in mind, and to consider requesting from existing groups.

CHAPTER 1

GROUND TRUTHING

FROM ISOLATION TO CONNECTION

YOU ARE HERE

This chapter invites you to learn these things:

- Your climate anxieties are a real, reasonable, and common response to the scale of the crisis we face. You can practice feeling them, sharing them, and learning from them without letting them destroy you.
- Current power structures want us to think that our individual, household-level actions are our only site of power, and that only governments (or companies) can make changes that matter. But neither of those statements is true.
- If you've lived through climate trauma, you have skills to share with people who haven't yet.
- We have many collective paths to meeting climate change, and many strengths and wisdoms we can put into action.

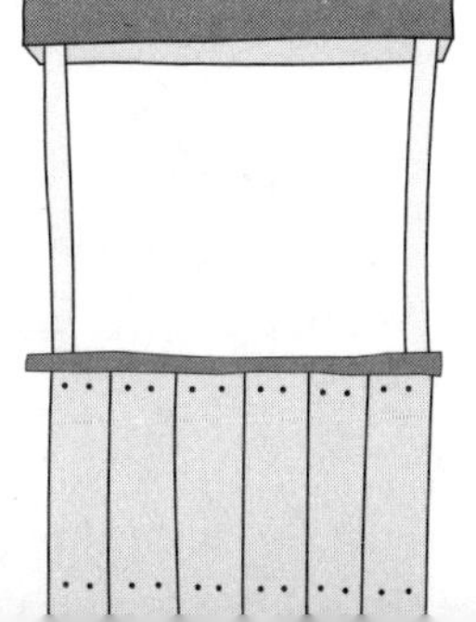

"I'M AFRAID THAT WE'RE LOSING"

Nicole Hernandez Hammer was already an environmental scientist when climate change came home to her. It came bubbling up through a Miami street on a sunny day in 2009.

"I was standing out there," she told me, "not a cloud in the sky, and I was on the corner of Alton and 10th, and the seawater just started rising up through the storm drains. And it was like a horror film. I felt so insignificant and futile and powerless." If the street was flooding with no rain, what would happen in hurricane season? The seawater rising in Miami was both a threat to the livable city—a setup for fried electrical systems, collapsing buildings, and people cut off from work or hospitals or childcare by unnavigable roads—and an indication of seas rising, irrevocably, everywhere.

Nicole had lived in Florida for much of her life after moving from Guatemala as a child, and she'd been working in climate science for several years. She didn't need convincing that human activity is changing the Earth's climate and the living conditions for everything and everyone on the planet. But that wasn't the same as seeing the flood come up to meet her.

"I'm afraid that we're losing," Nicole said to me the day we met at the Climate Anxiety Counseling booth. "That we're going to lose this. In the movies, you always know that things are bad when the scientists are freaking out. But scientists *are* freaking out, and it feels like a lot of people don't care. . . . Miami is already flooding. California is already burning. You go about your business, you hang out with your friends, people come over for dinner—and then you check the weather report." And you see three storm warnings back to back, or another day too hot to take the kids outside to play.

For many of us, the effects of climate change are both a source of future fear *and* a frightening memory. Nicole told me how she and her family lost their home in Hurricane Andrew in Florida in 1992. During the storm, they sheltered in Nicole's mom's house, the grownups shielding the kids with couch cushions and their own bodies when the windows blew out and the door slammed open.

The hours of the storm live in Nicole's mind. Years later, they sharpened the significance of the seeping water in the streets of Miami; later still, they amplified the terror of a flash flood that swamped her car on a Florida highway. They lift her resolve to make change, to meet change. They swirl through her desire to have a safe, gentle, stable life for herself and her son.

LIVING IN THE AFTERMATH

When climate anxiety reaches a certain level among a group or society, it can become what psychologists call *collective trauma*: a weight on our spirits that people feel even when they don't consciously recognize the drastic change or loss that's causing it. Living through disaster or ongoing environmental injustice can produce another kind of trauma: a loss or displacement that deeply transforms you and makes you vulnerable. It can be isolating and lonely, not just hard to talk about but hard to communicate through.

If you feel these ways, it's understandable if you want them to stop. And most of us fear losing what we have, however much or little it is. But our fear and grief and anger are not just valid responses, they're important and powerful guides. Allowing yourself to feel them, respect them, move through them, and return to them deliberately is the first step in hearing what you're telling yourself, and what the world is telling you, about what you can do—not as "only one person," but as one among many.

When the reality of climate change hits you, instead of pushing it away, try the following.

Steadying Your Nerves

This practice, which you can use anytime, will be particularly useful to you if you react to anxiety with agitation and escalation. Caroline Contillo, who first shared it with me, is a disaster resilience social worker. She's preparing people and communities to meet the effects of climate change with less anxiety and more mutual care.

If you're not used to handling strong emotion by doing a deliberate, physical thing, this practice might feel strange and beside the point. But there's a physiological reason why it helps: both the breathing and the sound-making affect the parts of our nervous system that don't respond to our reasoning or our will, bringing our parasympathetic nervous system (which helps us calm down) into play to counter our sympathetic nervous system (which riles us up). The same is true of the other grounding and bodily practices you'll find in this book. You can answer these questions in your mind, say the answers out loud, or write them down; involving your body through writing or speaking will allow you to feel them more.

QUESTIONS

Do you worry about climate change?

What, specifically, do you worry about?

PRACTICE

After answering each question, take a deep breath and breathe out while humming a low, deep hum each time. If you can't hear, or can't make sounds, try rounding out your lips and blowing out steadily and hard.

If you're doing this practice with a group and can hear, listen to each other's humming sounds as well as your own, and the total sound you make together.

This exercise is a low-stakes way to practice attention to one another. And it can also give people you're with a chance to match your calm, instead of matching your distress. Somatic coach, grief worker, and educator Selin Nurgün emphasizes that we need each other to transform the way we relate to the world, because change happens through relationships. ("Somatic" covers our entire state of body, mind, and feeling—and because we're always in relation with other living beings, it includes each person's self *and* the people they're with.) When you are able to feel, physically, what you care most about, and align your choices and actions with that caring, you become a "walking invitation," Selin says, for other people to join you in both feeling and action. "Bodies scan and read other bodies, very intuitively, that's our evolution. We can't heal alone. That's the important part of this work."

And Caroline confirmed that a sense of safety and calm doesn't come from an absence of risks, because it's not possible for there to be no risks. Our sense of calm, and our ability to meet stress and danger, comes from a combination of physical responses, which we can learn and teach, and social support, which we can build and nourish.

IS ANYONE OUT THERE?

When Nicole and I first spoke, I was freshly struck with horror at her bleak and blunt vision. And I was also relieved because what she was saying was what I'd been feeling. I was not alone.

Britt Wray, author of *Generation Dread: Finding Purpose in an Age of Climate Crisis*, writes about bringing her own climate anxieties to online sessions of the Good Grief Network, a peer support network for processing the uncertainty and grief of the climate crisis: "For an initial meeting, things got very real, very fast. Inside the four corners of my laptop screen, a bunch of strangers' heads . . . shared stories of environmentally linked depression, anxiety, hopelessness, and, for one person, the belief that societal collapse was coming soon and they weren't going to be alive for much longer. . . . Hearing how difficult other people's emotions were helped me feel less alone in my own." She goes on to describe how this sharing process, guided by the Good Grief Network's ten steps, led her and her companions through a sense of shared responsibility within uncertainty to a potential course of action.

The Good Grief Network didn't exist in 2014 when I started the Climate Anxiety Counseling booth. But other research and recommendations on climate anxiety, like Kari Norgaard and Ron Reed's research and Renee Lertzman's work, and guided programs for connecting with others to process those emotions, like Joanna Macy's Work That Reconnects, had been around for years. Not only did I not know about them, I didn't even think to go looking for them. Nor did I look around (at first) for opportunities for climate activism in my city or state—people who could have guided me in putting my love for the living world to work, people who were already struggling with environmental damage that my whiteness and economic security had cushioned me from. I was sunk too far, not just in climate anxiety but in the myth of isolation.

Mainstream US society's emphasis on self-sufficiency tells us that we're supposed to be able to handle everything ourselves, to not need "handouts," that Americans "got where they are today" without help from anyone. These cultural stories lead us to blame ourselves and, sometimes lethally, each other when we're hurt or high or broke—or when our dream house turns out to be in a flood zone. Blame, like harm, travels down the hierarchies of power and status: the more you've been taught to think that someone's less because of who they are, the more you assume that their suffering comes from something they did. We do this instead of asking why *else* a bad thing might have happened and who else might *benefit* from having it happen. There are also psychological reasons for this way of thinking: if bad things only happen to people who do or think bad things, then we can control whether we suffer or not.

With these pressures, we often keep our feelings to ourselves and seek personal or household-level responses to climate change, both because they're what's been presented to us and because we can control them with *relatively* little effort and almost no vulnerability. Hanging out laundry instead of using the dryer does take work and time, but it's not emotionally risky, and it's less likely to require negotiation, confrontation, or even openness with other people. The next exercise will help you reflect on how you currently talk about climate change on a day-to-day level, and let you practice bringing it up in a relatively manageable way.

Saying It Out Loud

This practice was a gift from a young visitor in the booth's first season, who told me they started small climate conversations like this all the time. It combines something common to talk about—the weather—with something much harder to talk about. And it holds the reality that climate change is here and now.

QUESTIONS

Do you talk with people about climate change?

If you don't, why not?

If you've done this, how have people responded?

How did you feel then?

What did you say to them next?

PRACTICE

Next time someone mentions the weather (any person, any weather), bring up climate change. It might mean saying, "Yeah, I treasure every snowy day now," or, "I heard we can expect more rain like this as climate change gets worse," or, "My aunt was really hurting last week. I don't know what she's going to do if we get another heat wave." Say it in your regular words and voice, written or spoken. Notice how you feel physically when they respond, and pause before you react to them in a way that shows.

If you're doing this practice as part of an intentional group, you can act out these exchanges with each other before trying them elsewhere, and let each other know how they went.

When I started offering Climate Anxiety Counseling, I was feeling climate change most in the destruction of ecosystems that aren't my home and that I'll never visit. Many people who come to the booth think of it that way, too: in our New England city center, they talk about polar bears and island nations. But all our places are vulnerable to climate change and the forces that cause it. Where I'm living, Narragansett and Pocasset Wampanoag people have spoken publicly of the new threat that sea level rise and warming oceans pose to the marshes where they fish, hunt, and quahog, and where both food and history are nurtured. In my role as their neighbor as well as a settler on their land, listening is the least of what I owe them. Listening reminds me of the vulnerability of this place that has become my home. It also reminds me that the same landscape is haunted by different losses, and that I need to care for so much more than what I'm personally losing.

All of us are part of living systems of sustenance much bigger than we are. We're also part of the economic and political systems that keep climate change on the increase, while already hurting us in many other ways. Climate change presents us with the challenge of where to put our energy and effort in order to change those systems—and to do that, we have to talk to each other about it.

ACROSS DIFFERENCE AND INDIFFERENCE

At the beginning of Climate Anxiety Counseling, I'd get frustrated or even enraged when people's reactions to climate change didn't match my own. I've never stopped struggling with this, in all my years of "doing the booth." But I learned that I had to listen to the reality of what people were seeing and feeling before I had any chance of working with them to shift that reality.

That was especially true for the many people who stopped to talk not about climate anxiety but about their other anxieties, pressures, and needs: tooth pain they couldn't afford to treat, racist customers at work, the struggle to get support for a child with learning disabilities. One woman was headed for south Florida, right into the floods and the storms, because in Rhode Island she and her two small sons were living in a homeless shelter. Many said that climate change did worry them, but not as much as their more immediate needs. Their pain was clear—and so was their point.

If you recognize elements of yourself and your life in those portraits, you're not alone—and as you may know already, these aren't just personal problems. These anxieties of health, housing, and education have the same roots as climate change: economic and political systems, or structures, that rely on inequality, domination, and the extraction of work and material. Systems *contain* and *direct* people—we are parts of them. The two main systems that drive climate change are capitalism, which chews lives up to create value, and white supremacy, which claims that doing so is okay if the lives are not the lives of white humans. While these systems affect people differently, and drive us apart from each other by doing so, when we talk together across those differences we lay the groundwork for greater fairness and more lasting well-being in the ways we work, eat, find and maintain shelter, and deal with one another.

The year Nicole Hernandez Hammer saw the water come up through the street, she also saw a national list of the places in the United States most vulnerable to climate change. "Actually to sea level rise specifically. I realized they were a lot of the same places I had family or friends—California, Texas, New York, Florida, and some others. And so I pulled up the census information and, as is kind of obvious, those places have some of the largest and/or the fastest-growing Latino populations in the country. I felt like I was in a unique position as a Guatemalan immigrant and a Spanish-speaker *and* a scientist to be able to start talking about it."

She quit her job and placed her scientific skills in the service of two nonprofits, Moms Clean Air Force and the Union of Concerned Scientists, "putting together different groups, doing town halls, doing meetings, doing outreach. And then I developed a model: when is the climate change impact happening? Where is it happening? Not when it'll happen in a hundred years or ten years, but when next year will I be most likely to see the evidence of climate change impacts, and where? Not, like, what country, but what intersection?"

When she found those intersections, those exact streets—climate-sensitive places like those in south Florida—Nicole would talk with people living or working near there. They told her about cars whose waterlogged engines wouldn't start, keeping them out of work and their kids out of daycare, and cuts infected by bacteria from backed-up sewage drains. These conversations led to meetings with local politicians and connections with other local justice groups and coverage from nationally recognized journalists and, eventually, changes in climate policy for both the state and the nation.

These kinds of large-scale changes take a long time, and a lot of relationships, to build—they took Nicole and her research and community partners several years. But you can start by assessing the potential for politically driven change right where you are.

Developing Climate Civics

I'll be honest: this is also an exercise to identify what's *not* happening, insufficient, or continuing an unjust pattern. Before we can evaluate laws and policies, and choose which to support, oppose, or work around, we need to build up a picture of what's being done at the government level, and what's being ignored or dismissed.

QUESTIONS

What are the laws and policies in place, in your state or city/town, that limit, or provide for, the effects of climate change?

What are the laws and policies to do those things that have been proposed but aren't happening yet?

PRACTICE

If you didn't know those answers, this is a chance to look them up. If you're doing your own internet search, try typing this into the search bar:

- Site:gov for US government sites, site:canada.gov for Canadian government sites (this filters out a lot of junk)

- The name of your city/town
- The name of your state/province
- One policy word, like "regulations"
- One keyword or phrase, like "emissions"

Some useful keywords include "environmental justice," "emissions," "restoration," "green infrastructure"; some useful policy words include "legislation," "bill," "regulations," "ordinance." Or ask a public librarian, in person or online, for guidance in your search. Make notes or use browser bookmarks to keep track of what you learn.

If you're doing this with a group, gathering to do research together can be fun, especially with snacks. You can divide up search terms, offices and agencies, areas of interest, or search for legislation or policy that's related to your own fields or industries.

Because climate change is huge, putting the huge machinery of government in motion to respond to it can be effective *if* it actually happens: restricting fossil fuel mining and burning, and subsidizing new jobs for people who currently do related work; removing laws that trap people in threatened areas, and coordinating and funding the relocation of communities threatened by climate disaster; investing in the health of ecosystems by changing who directs what happens there.

But governments and the people invested in them are also often reluctant to change the structures that got them where they are. They too are moved by fear of losing what they have. Their fear and delay, temporarily profitable for them, are deadly for others.

In panic at losing tax revenues from pricey coastal properties, one of Miami's responses to sea level rise and flooding has been to swiftly permit luxury development in Black and Latinx neighborhoods on higher ground: longtime residents are not just pushed out of their homes, but driven onto lower, more frequently flooded lands. Lobbied by industrial growers and developers, California lawmakers have rejected bills to make water use more equitable or regulate rebuilding in fire zones. And because the governor and the energy company refused to upgrade Texas's power grid, Texas residents lost homes and lives in a climate-change-induced freeze for the third winter in a row. That level of obstruction—what futurist Alex Steffen calls "predatory delay"—is a major source of the helplessness, anger, and sorrow that I feel, that people express to me at the counseling booth, and that may be one reason you picked up this book.

It's in the interest of capitalism and white supremacy (and the people they enable) for us to think that our individual, household-level actions are our *only* site of power. It's equally in their interest for us to think that *only* governments (or companies) can make changes that matter. Both trap us in isolation and passivity. It's through conversations—at the booth, in line at the post office, with friends over dinner, and online—that I started to find and feel the power of what people can do together even when governments refuse to step up.

BUILDING EMOTIONAL RESILIENCE AND PRACTICAL POWER

When Superstorm Sandy ripped through Brooklyn and the Rockaways and New Jersey and lower Manhattan in the cold fall of

2012, years of predatory delay—neglect and bad maintenance in streets, subways, electrical wires, and buildings—met the force of wind and water, and collapsed. Poor people redlined out to the coasts and low-lying areas were hardest hit, and once the storm subsided, they were left longest with its damage.

News media referred to Occupy Sandy as a "rag-tag team," but in fact Nicole Capobianco and the rest of the autonomous emergency response crew were better prepared than many of the people and agencies whose job it was. The complicated but mutually supportive and joyous encampment at Occupy Wall Street, where the crew first knew each other, had taught them how to collectively meet people's needs. "So it was Halloween, and instead of a trick or treat we did a trick or donate type thing. We got on a conference call and some people were willing to be these little hubs for supplies, and some people were willing to gather the supplies, and then we figured out who has a vehicle to *move* supplies, 'cause the subways were compromised, and we just started compiling the various resources that we had." She went door to door in Brooklyn ahead of the storm, and residents brought out what they could spare: blankets, powdered milk.

"Then the storm actually hit." The power went out. The elevators were down. "So we started to set up these groups of people who were willing to climb the stairwells with supplies, check on elderly people, check on disabled people, check on people with children. And, like, horrible stories. My friend found somebody passed away, like—there's—it was—we were doing what the Red Cross was not willing to do. FEMA was handing out peanut butter sandwiches. It was a joke."

After a deep breath, she emphasized that it was because the crew already knew each other and had practiced working together that they were able to mobilize so quickly. She also thinks that New Yorkers reacted better to Occupy Sandy than to Occupy Wall Street "maybe because it was seen as more nonthreatening as an aid organization than as, 'Hey, Wall Street is responsible for climate change,' even though that would have been our political stance if you had asked. But it was also about developing mutual aid networks. And it's about what we're able to accomplish when we allow people to contribute what they're best at, and you get to see the way that different people shine in different situations, providing all kinds of different care."

Take a moment, and a breath, to revisit your fears for a chaotic future—or your memories of the recent past. Do one of the steadying or grounding practices again if you need to, or even if you don't: making them a habit is helpful. Then continue to this next exercise, which is similar to Nicole's work with Occupy Sandy and gives you at least one way to respond to your fear.

Getting Ready and Getting Rested

If climate disaster comes to your home, this could be how you meet it. It works both ways: recognizing your reactions to climate change helps you to meet it with care, and having a plan for meeting climate change with care (and with others) reduces the need to push your reactions away.

QUESTIONS

Where will you start, after the next disaster?

Who and what can help you make room in your life—for more learning, and also for rest?

PRACTICE

State and city emergency management associations sometimes offer free trainings to join a Community Emergency Response Team (CERT). Some cities also have Communities Responding to Extreme Weather (CREW) teams. Research these, and if there's one near where you live, note the locations and dates and times of the trainings (most are once a week for about six weeks). Make a short list of people you know well, who live nearby, and who you could invite to attend a training with you—and people who could help you make the time and find the energy to attend and learn. If in doing this you realize that what you really need is to use that time for rest rather than doing the training, ask for support in that, either from the group you're doing these exercises with or from other people in your life: evenings of rest once a week for six weeks. Then see what you find yourself ready to do.

This is a good time to check in with yourself: How does it feel to read that practice? Committing to a training—not to mention doing the thing you're training for—raises challenges of time, energy, and attention. If we want to add more effort to our lives, we also need to integrate rest into our effort from the very beginning. If you have kids and so does someone else who wants to do the training, maybe one of you attends and shares the information and skills later, and the other puts on a movie and fixes dinner for both families. Balances and exchanges like these open an *ongoing* practice of rest, mutual support, and reciprocity that runs through this book—and through the transformations that we need in our world.

Throughout the COVID summer and fall of 2020, eight years after Superstorm Sandy and with the marks of the storm and the city's neglect still on the neighborhood, Tiffiney Davis mobilized residents, food distributors, stores, and restaurants of Red Hook, Brooklyn, to feed their neighbors through the tightest times. Tiffiney is the executive director of the Red Hook Art Project, originally a Saturday art class for middle and high school students. With her vision and direction, it's become much more.

Growing up and raising her two kids, Tiffiney had to become an expert in the New York city and state systems that controlled her food and shelter. Now, when a student at RHAP trusts her with the news that they have lost their apartment or have a violent person in their family, Tiffiney knows exactly which numbers to call and how long they need to spend in the shelter to qualify for the right waiting list. She marshals neighbors to get them food and grocery gift cards. She makes sure the grownups in the family know that they can demand to be sheltered forty-five minutes or less from where the children are in school: "It is *your right* to remain an hour or less away from your support system." And she respects, and learns from, people's own knowledge of what they need.

The mutual aid and food distribution she coordinated during COVID lockdown was an outgrowth not just of her systemic expertise, but of her long relationships with Red Hook families. And while RHAP students distributed lunch bags and local restaurants stocked the community fridge with burritos, Tiffiney began to solidify a long-held dream of a real community hub. She envisioned a clean and safe space to cook and eat together, where the gas and the electricity stay on, and people can learn from each other. "When the people of Red Hook are not hungry," she wrote in a grant application, "their minds and bodies will be strong enough for healthy conversations, healthy relationships, better sleep. Free to focus on education and think about the future. When you are healthy and safe, you can use your voice."

Climate change is both adding to and causing disasters that are longer, lifelong, generations long. Acknowledging both our limits and our strengths is key to choosing our course of action and the best places for our effort. Here's an exercise for doing what Tiffiney did: drawing on your own experience of coming *through* difficulty, with witness and assistance from others, to help you feel capable and strong.

Drawing from the Well

This exercise is part of the "social support" that Caroline Contillo mentioned as a root of our sense of safety: a reminder of past connections and a guide for connections to come.

QUESTIONS

What disasters have shaped your life?

How have you cared for yourself and others within those disasters?

PRACTICE

Without speaking, bring forward a memory you have of safety and connection with other people. Build it up one detail at a time, making the memory clear and intense. If this makes you sad or lonely, because of relationships that have ended or changed, you can thank those relationships for being what they were—even if you no longer want to thank the people themselves.

Next, bring forward your memories of yourself in a time when you have reached safe harbor, even for a short while. Build up your picture of yourself at that time, making it clear and intense. If this makes you embarrassed or impatient, you can thank your past self for being who you needed to be to reach that harbor and benefit from it.

Now, take any safety and confidence that has come from bringing these memories forward. Ask it to be ready for you when it's time for things to change again.

If you're doing this exercise as part of a group, you can share and witness each other's memories, or just accompany each other in silence.

Wrangling policy, running emergency redistribution, building a community hub, even starting public conversations like I did at the counseling booth—maybe you're looking at the people you've met in this chapter and thinking, "If I could do something like that, I'd already be doing it." But before Tiffiney, Nicole, and Nicole started meeting short and long disasters in these ways, they *were* doing something else. They recognized what they saw and felt themselves, and what it meant about what their communities needed. They took the leap out of isolation and learned from people, circumstances, and systems around them. They drew on the skills, relationships, and knowledge they had built over time, and brought the pieces together to be part of something necessary and new. Like them, you can change to meet the future in a way that's rooted in who you have been, and your connections with others will help you move into who you can be.

BECOMING MORE COURAGEOUS—AND MORE VULNERABLE

When the storm comes, you move from the front porch or the stoop to the entryway. When the bedroom windows break, you move into the hallway. And when the wind starts to blow through the broken windows against the bedroom door, you move from the hallway into the living room. You move closer together, close to the center of the house, the place farthest from the storm. And then you pile the couch cushions on top of the kids, and hold them down with your body, and if you believe in God, maybe you pray.

With climate change, you might be the kid in the middle of the pile, the parent trying to shelter them or struggling to wedge the front door with a table, the absent parent on the other end of the phone hearing the sound of wind and screaming. Or you might be reading about it in the news. Those of us who don't have practice in change and transformation can learn from those of us who do. This is how isolation ceases to control us.

Everything that has happened to you belongs to you and is yours to share. If you do share it, it can help not just you but all of us survive. Here's a practice for identifying how much you have to offer. In thinking about what you have to offer in a crisis, you also learn what you'll need to reach out for from another person, group of people, or body of knowledge. When we admit what we don't know, we offer someone else the chance to teach us.

Taking Inventory

Selin Nurgün encourages friends, accomplices, and clients to give gratitude even to the reactions that we might want to move away from, because at one point, they offered us something we needed: "What is this protective strategy? What is it doing for me and what is it taking care of? Is it still useful to me? I'm a big fan of saying, *Thank you, you did so great. You brought me here.*" You can add that gratitude to this exercise if you like.

QUESTIONS

If your fear and survival have given you strengths, what's one that is useful in the kinds of difficulty and pain that climate change is bringing?

What's one that is useful in a short, sharp crisis?

What's one that is useful in a disaster that drags on?

What's one that is useful in an emergency that ebbs and flows?

Which do you want to share with others?

PRACTICE

On a piece of paper, on your phone's notes app, or using voice-to-text, write down a disaster—short or long—specific to where you live. You can choose one that has happened, is happening, or is likely to happen, where some effect of climate change is a factor, and where *one* of your skills could come into play. Around the name of the disaster and the thing that you could do, write what other people, groups, and institutions would need to do, or could do, to increase survival and well-being for the living beings in that place.

If you're doing this as part of a group, share. Whose skills complement one another? Whose skills are similar or overlap, so that each person who has them could rest sometimes? What could you learn from each other?

How you feel about climate change matters—not just to you, but to the people around you—because it can move you to respond to climate change together. When fear flings us into the past or the future, we can stand together in the present and steady one another, not because we are safe but because finding calm and staying together may help to bring us through.

CHAPTER 2

"THE WORLD BELONGS TO EVERYBODY"

FROM FEAR AND HOARDING TO PLANNING AND WELCOME

YOU ARE HERE

This chapter invites you to learn these things:

- You can use your fears about the climate crisis as a starting point for learning and meeting your community's needs as well as your own.
- You can survive climate impacts without aligning yourself with forces of domination, control, and punishment.
- You and your community are already rich in the skills, materials, and relationships that will help you live within climate change.

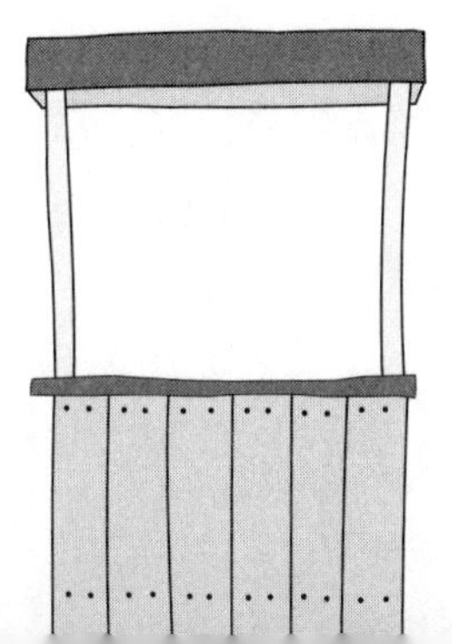

"DO YOU HAVE ENOUGH SPACE?"

In July of 2015, I talked at the booth with S, a young Cambodian American social services worker. S did feel climate anxiety, but spent most of our time together talking about the three generations of family crowded into her house out of economic necessity: the limited space, the noise, the amount of time they spent fighting and frustrated with one another. "Because it's summer, with the kids being outside so much, the noise factor has dropped. But during the winter I couldn't breathe, I was having chest pains. It was major. It was every day." S spent a lot of time at her boyfriend's house to try to get away. "People will say, kick 'em out, but..." She trailed off. "It was supposed to be temporary, and then after two years I was like, this is what's happening."

S came back to see me in 2017. "My life has changed," she told me. "My brother moved out, so now upstairs is my mom, my boyfriend and me, and downstairs is my sister and her kids, and it's so much better, because we all have enough space. Now whenever I meet someone who's going through it, the first question I ask is, 'Do you have enough space?'"

When disaster or necessity strikes, we call on the people we're closest with to take us in. Yet so much of what people say to me at the Climate Anxiety Counseling booth already includes the words *no space* and *no time*: parents staring down a food-smeared kitchen after yet another bedtime struggle and before another day of work. Workers trying to maintain their jobs while managing chronic pain. People already striving to address other injustices, in housing or education or health, unable to get their colleagues to make the connection between that work and environmental justice. I realized that I was hearing *fear*: fear of being eaten away, encroached on, eroded by even more demands. Just seeing my Climate Anxiety Counseling sign prompted people to feel that they should be "doing more" to respond to climate change, followed quickly by: *But don't ask me for anything more. I have nothing to spare.*

Figuring out whether that's true for you, and what would change it, takes room and time in itself. It's not enough to know what's right and wrong; we also need to feel that what's right is possible.

At the booth, people also worry aloud about protecting themselves and their families—an understandable and powerful anxiety. When they talk about doing that by stockpiling food, buying gold, building fences, and amassing guns, I'm saddened but not surprised. The myth of isolation outlined in Chapter 1 can harden into contempt, and blame, and fear.

By recognizing that form of fear in ourselves and others, we can shift away from it toward a more resilient and welcoming way of being. This next exercise will help you identify and source your fears without getting stuck in them.

Moving Through Fear, in Place

While we can reflect on fear with these questions, and our thoughts can bring it on, fear isn't a thinking emotion, but a reaction. This practice of tension, relaxation, and motion is a way to process any fear that you notice yourself feeling, and you can use it throughout the book or any time you want.

QUESTIONS

When you imagine staying in the place where you live now while the climate changes, what's one thing you fear?

When you imagine having to leave the place where you live now because of climate change, what's one thing you fear?

PRACTICE

After you name each fear, tense and then relax everything in your body that you have the ability to tense and relax. If that's likely to cause problems for you, vividly imagine rigidity and then looseness in a way you know won't hurt you.

Then do something physical that uses the full range of motion for how your body is at this moment—you know best what that is. If it helps, visualize whatever fears came up in the first part of the practice washing out of your body as sweat or water.

If you're doing this practice with other people, tense and relax for their fears as well as yours. You may want to do it with eyes closed and/or backs turned so you're not worried about how your face or body appear.

You'll notice that I didn't ask you whether the fears you brought up were realistic or well-founded or likely to happen. That's because the mental and emotional map of fear is also a map of the power we *perceive* ourselves to have. It's not identical with a map of what's actually going to happen or what we're going to do.

And recognizing that—placing ourselves on that map of power and fear—is essential to being a purposeful and caring part of a climate-changed world.

UNCLENCHING OUR HANDS IN THE FACE OF DISASTER

Eva Amanda Agudelo came up to the Climate Anxiety Counseling booth in May of 2018. We'd met before: she, her carpenter husband Mattie, and their three-year-old had held a banner with me at a windy rally against a liquefied natural gas plant. Mattie was with her this time, too, and their conversation was mostly with each other. "I'm worried that buying a house in Rhode Island was a terrible idea because of sea level rise," she said.

"I checked," Mattie said. "Our house is seventy-five to eighty feet above sea level."

This raised Amanda's ire. "But what about all the people who aren't seventy-five to eighty feet above sea level? You can't live in a world where your neighbors are flooded out and you're fine. And then am I gonna have people breaking in because they're starving? We can't survive unless all of us survive."

A lot of people get as far as *Am I gonna have people breaking in because they're starving?* but never move through that fear to the next question of shared survival. "We've been planting food in our yard," Amanda went on. "The solution to scarcity is to offer freely, so you have to become a producer and have something to offer. I can't feed everybody, but maybe I can feed people enough to keep someone from hurting me."

There's still fear in that last sentence: the fear that jumps ahead to a future where other people will of course be the enemy.

Notice—gently, without judgment—how you hear it, whether and where you recognize yourself in it. Who do you think you'd be in that scenario, and why?

Catching Yourself

This exercise will help you zero in on where your fears come from—and where they tend to lead you. It's especially useful for the doomscrollers among us. Remember that a sense of safety doesn't come from an absence of risk; remember, too, that a risk doesn't need to be real or close for it to disrupt your sense of safety. Nor do you need to be right about its source, or whose fault it is, to feel fear. Our goal is not to diminish what has already happened to us or push our fears away, but to make a separation between feeling them and acting on them, so that we have time to figure out which course of action would be best.

QUESTIONS

Which of your climate fears are based in your experiences?

Which are based in media or reporting you've seen yourself?

Which are based in conversations you've had with other people?

PRACTICE

Watch, read, or listen to the news, or scroll your social media feed(s)—whatever routine you have for taking in news about the wider world. Starting now, each time you do this, notice and record (in some way that's convenient for you) when you feel weak and when you feel strong, and what you were seeing or hearing that brought on those reactions.

Then, start noticing and recording *who* is speaking, how they're describing or labeling themselves and anyone else they're talking about, whether they're recommending a course of action, and what it is. Be as neutral as you can in your descriptions—"Seeing X confront Y in that video made me feel strong"—and use as little self-critical language as possible. Do this for at least a week and note any changes in the ways you react or the things you react to.

If you can, check in at the end of the week with someone in your group who's doing this practice. Describe to each other what you noticed—and pay attention to how you react to their description, too.

In our conversation at the counseling booth, Amanda put into words the feeling you get when you know that things are already unfair and that climate change is making them more unfair, and you're struggling with what that means for you and the people you love. Many of us are all too ready to look for reasons why community responsibility isn't (or shouldn't have to be) the same as protection for our families and ourselves. Sometimes that's because our communities have failed us in the past. But to build trust and power together, we first need to let go of any illusion that we owe each other nothing.

If we carry that attitude into climate change, it will kill us, and it will kill first those of us who are already suffering worst.

UK writing and thinking collective Out of the Woods, who study climate change and social control, put it very simply: "Climate change is another reason to have to move, but it is not a reason for states to treat moving, racialised people any differently" than they have in the past: with hoarding, gatekeeping, violence, and blame. *They should have stayed in their country, their city, their neighborhood. They brought it on themselves.*

The name for this attitude, and the activity and policy it shapes, is ecofascism: dictatorial and punishing control of people's lives and movements with "the environment" as an excuse. In and at the edges of the United States and Europe especially, this attitude directs immigration policy, food and housing access, disability assistance, and other government actions that will affect more people as the climate changes. It also leads to harsher state punishments for attempts to stop environmental damage, like the violent arrests of people blocking the Line 3 pipeline on Anishinaabe lands. You can hear ecofascism in claims about "overpopulation," "pristine" or "untouched" landscapes, "burdens on society," and who "deserves" to live, to be healthy, or to inhabit land. In environmental campaigns and movements, it shows up in the assumption that people with, say, more college education know more about what's best for the land and its people than those who live there, or that more top-down control and punishment of people's small-scale activities is the way to "save the planet."

Ecofascist attitudes or policies won't save even the people or places they claim to be defending. They will destroy a lot of other people and places, and they will distract us from things we *can* do to respond to climate change more thoroughly and fairly. Being compassionate to yourself about what you have felt and honest about what you have done are both essential to doing differently. If you find yourself drawn toward these controlling, fear-motivated positions, or hear other people embracing them, invite yourself—and them, if you can—into the "Moving Through Fear, in Place" exercise from page 27.

That practice helps us live honestly with the fear (and secondary emotions like anger and shame) that can be provoked by our knowledge of climate change and inequality, emotions we might try to avoid feeling, push down, or justify. If we push them down, they don't go away—they can show up in our actions and choices, what we approve of and what we let go by. In a racist culture, they'll target racialized people. In a culture that maintains contempt for poverty, they'll target impoverished people.

After Hurricane Ida in 2021, New Orleans residents reported police stationed outside a Whole Foods supermarket. Presumably both the police and the store managers intended to let the perishable food rot in the cases rather than let people cook and eat it without paying. Both cops and store managers were choosing to hang on to their power in a moment when making a different choice could have helped people survive. Instead, power, fear, and blame all stayed where they were before the storm, and hungry people in New Orleans still didn't have enough to eat.

That story isn't just about what the police did: it's about people who *aren't* police (which is most of us) and what we're willing, or even eager, to have police or other authorities do when we are afraid. It's possible—likely, even—that there were storekeepers and store workers in the city who overtly or quietly shared what they had before it went bad. But that didn't make the news, and that matters, too. We don't have to affirm or reinforce or praise the decision to hoard nourishing food, livable shelter, clean water, whether at the community level or the government level. At ecofascism's heart is the question of who deserves to live. But this is not really a question: it is a trap.

If you've fallen for it, even a little, feel your embarrassment and let it go. Then turn to some real questions you can ask yourself and your community.

Turning It Around

This exercise works best with people who perceive you as being like them in important ways. Insisting on conversations like these, as earnestly as possible and with as many people as possible, is how we identify the abundance, strength, flexibility, and welcome that already exist in our communities, and activate the much more that we are capable of.

QUESTIONS

How will we make room for everyone who needs housing?

How will we find food for everyone who needs food?

PRACTICE

When you hear people you know asking questions like these in a way that secretly means "We can't possibly...," respond as though they're asking questions whose answers they really want. Invite them to strategize with you about those answers. One way is to agree with them about the *level* of difficulty ("Yeah, it's going to demand a lot of all of us") and then step quickly into *addressing* the difficulty together ("Do you know if such-and-such church has any kind of mentorship or sponsorship program?" or, "You know, the houses on so-and-so street have been abandoned for a long time . . ."). Of course, use your own regular voice and words. With a little practice, you can do this even when someone makes a statement—"We don't have enough"—rather than asking a question. Be open, be warm, and keep them going as long as you can.

If you're doing this practice as part of a group, you can role-play a few of these for when you encounter them elsewhere.

FINDING OUR LONG-TERM STRENGTHS AS A COMMUNITY

Not long after our conversation at the counseling booth, Eva Amanda Agudelo came up with a partial answer to the aforementioned question—*How will we feed everyone who needs food?*—by leaving her job at the RI Food Bank to found Hope's Harvest RI, a food rescue and distribution nonprofit. Hope's Harvest coordinates volunteers to gather fruits and vegetables that participating farms don't think they can sell to their usual customers, and distributes them to hunger-relief agencies: about 48 separate ones, plus the food bank's network of over 150 more. Throughout the first two years of the COVID-19 pandemic, they delivered 505,353 pounds of fruits and vegetables.

In Hope's Harvest's second summer, Amanda took me with her to pick up and distribute a donation. At the Olneyville Food Center, we sorted through pallets of vegetables, extracting moldy squash, yellowing kale, and cucumbers with slimy spots. "What I learned from working at a food co-op and then at the

Food Bank," Amanda said, "is: if you wouldn't eat it, you can't ask anyone else to." Olneyville is one of the main places you live in Providence if you've been displaced from elsewhere—including by climate-change-fueled flooding, fire, and drought, and by climate-changing industries like mining, logging, and industrial farming.

There's an elegance to food rescue, a satisfaction. Hungry people get to eat; the food gets to *be* food, rather than waste. Farmers get paid for produce that they otherwise couldn't sell, and volunteers learn more about where their food comes from and who grows it. Hope's Harvest and its network of partners do work as various as convincing foundations to support the project, stooping in the hot sun, and planning the season for extreme and unpredictable weather—one of climate change's biggest threats to our food supply.

"Since I plant crops, I can see," Rhode Island farmer Blia Moua said when I told him I was writing about climate change and asked what he'd noticed about the growing season. "Global warming makes the climate *change*. I think we should do something about it. The world belongs to everybody. Each and every one of us." Blia's stand is on the other side of the library lawn from the Climate Anxiety Counseling booth at the Sankofa Market, where I've been "doing the booth" since 2015, among tables of vegetables grown and products made by people who live, farm, and work in and around Providence. Many of the vendors have roots in Africa or Southeast Asia, and the foods they grow provide whole neighborhoods with the taste of home. As part of the West Elmwood Housing Development Corporation, Sankofa stabilizes prices and doubles the value of SNAP to buy produce: affordability is also a factor. People come back every week to get honey and peppers from Teo, callaloo and sweet potato greens from Charlotte, and bitter melon and beans from Blia.

Between helping one customer sign off on her SNAP check and confirming for another that, yes, the cooking greens and white-streaked gourds are all organic, Blia told me that he'd been farming in Providence since 1989, soon after he arrived in the United States as a refugee. If you're new to a place, he said, you can't be sure whether what you're seeing is normal. But if you stay somewhere for a long time—and especially if you grow food there—you can tell the difference between the variation from year to year, and the more extreme changes that persist.

I wanted to hear from Blia as a farmer, and also as someone who had to make a life in a new place. "I was a little soldier," he told me: he protected downed American pilots during the Vietnam War. This allegiance got him into a refugee camp in Thailand when the war ended, and then into the United States. "It's hard, it's hard when you move one place to another. It's not the place that you move when you want to go. But you say, 'I don't know. I don't know what's worse gonna be. Is it going to be bad or good?'"

Climate change brings that uncertainty to more and more of us: *Will we have to go? Where will we go? What will it be like? Who else will be there? Will they make us welcome?* When I asked Blia what he'd tell people who have to make a transition similar to his, I learned that he already does this all the time. He's sponsored cousins to come from Laos, walking them through survival needs like finding a doctor and avoiding racist neighborhoods, and also through pleasures, like visiting national parks or the shore. He was the first from his family to arrive and had nobody to teach him. "It took me like twenty years to learn something, and I'm being sponsor to my family when they get here. Because I want them to *know*. How to get a job. How to save money. How to go to the bank."

If you've had to make your home elsewhere, as Blia did, you can be a guide for family or friends who are doing it for the first time. And if you have a lasting familiarity with the place where you now live, you have other kinds of knowledge and skills to share. Finding these in yourself, and finding other people whose knowledge and skills complement yours, is the next step on your path of meeting the anxiety of climate displacement. This exercise will help.

Declaring Your Abundance

Mutual aid—identifying and sharing what you can spare, people to people—is the antidote to hoarding. What you locate through this exercise can build a place that can accommodate anyone, whether they're passing through or have come to stay.

QUESTIONS

What administrative systems do you wrestle with regularly and successfully? (Insurance, immigration, unemployment...)

What materials are easy for you to get hold of? (Construction, garden, repair...)

What habits and practices of caring for others are familiar and even fun for you? (Elder care, childcare, tending plants or pets, cooking for people...)

What are you good at building, fixing, maintaining, or taking apart? (Cars, boats, bikes, buildings, blenders...)

Are you good at showing other people how to do something? At coordinating a bunch of people to do something together? At solving problems or resolving conflicts?

Which of these things are you *not* good at or comfortable with—but you know someone who is?

PRACTICE

Reach out to *one* other person you know to explore what you or they can freely share, either through a trade—their abundance for yours—or something you can share together that will eventually come back around and enrich the community. Accept something from them, or offer them something.

If you're doing this as part of a group, you can also help people you know identify their abundances—things they might not value in themselves but that you value in them. And for fun, you can wear all the abundances you're tapped into: while you're talking, have ready some little circles of paper, colored pencils or markers, and safety pins or tape. If they're able, each person in the group will draw and bestow a "merit badge" on one other person for one of the things they can offer. Looking around the circle will fill in your picture of what your neighborhood or community contains—as well as areas where you'd need to reach beyond your circle for expertise, resources, or even time.

Mutual aid can be as coordinated and complex as the distributions of food and toiletries that many communities muster for their houseless members, or as seemingly small as free childcare or repairs. We can recognize it by the way it maintains people's dignity, offers alternatives (even brief ones) to harmful systems, and lays groundwork for more shared efforts to survive.

WEAVING NETWORKS OF CARE

In the Climate Anxiety Counseling booth's first season, four people I already knew came downtown to visit me: a couple and two friends, all white and in their fifties. Like a lot of people thinking and worrying about climate change, they swayed back and forth between abstract causes and concrete effects, struggling to connect them. "The culture doesn't provide us with an opportunity or a venue or a language for what it means to be afraid," M said. "People think global warming means warming generally, not just greater severity. But what I noticed this year is that a lot of things popped up, came into flower, then blackened and died."

"Do you imagine the future?" I asked.

"The future is here," M's partner, J, said. "I don't have to imagine it. There are extreme environmental events right now that are happening to people I know. The flood in Boulder—cataclysmic." She was talking about the fall 2013 floods in Colorado, a climate-spiked disaster that killed at least eight people. "Nothing like that had happened during my lifetime."

I asked how they imagined they'd help and sustain other people in an acute crisis. "Well, on a really concrete level we would shelter people if we had to," M said, hugging her friend, C. "We have a lot of room."

"That assumes we'll be the ones who *have* the shelter," J pointed out.

"Well, it's fair to assume we'll be more likely to have it than some other people."

They went back and forth like longtime lovers, which they are, challenging each other to put together a shared understanding. "That's true," J conceded. "People who are on the coast, who are on the floodplain—"

"I was thinking more of people in island nations," C said, still from inside the hug. "My building would probably be fucked, but I'm on the third floor. I'll have to go in and out through the window."

"It shouldn't be about individuals sharing what they have," said J firmly. "It should be about changing structures so that nobody has to rely on somebody else's goodwill." It came through as we talked that she was drawing on her memory of the AIDS crisis, which all four of them had lived through while loving people who did not. The responsibility to *demand* that larger structures and institutions meet a crisis, and the responsibility to step up and offer care when they don't, are lessons that these friends and their friendship have brought with them into the climate present.

And why not both? Why not build your sharing out into a structure? Why not create structures that *embody* goodwill—unlike so many of our social welfare systems now, which also take a hoarding and gatekeeping approach? (A good half of the people who talk to me at the counseling booth say their anxieties would be alleviated if they could access social services—housing, food stamps, medical care, harm reduction, mental health care.)

A disaster or other major transition can—and often does—intensify gatekeeping and hoarding. Ask anyone who applied for

FEMA aid or insurance payments after hurricanes in Louisiana or fires in California. A major transition can also offer opportunities for systems to reorder themselves, even if temporarily, in response to community needs: compare the New Orleans Whole Foods with grocery chain H-E-B, whose disaster response team includes community kitchens, clean water tanks, and check-cashing and prescription-filling services. They mustered in response to Hurricane Harvey in Houston, lining up trucks on the raised HOV lanes, ready to roll in when the rain stopped—and roll in they did. That's not the total restructuring of systems that J was calling for. But it's a taste, a hint of what's possible *besides* what happens—and hurts us—most often.

Mutual aid is good to practice whether you're in a disaster or not; it builds up a habit of sharing that you can call on when the need is more urgent. The same is true for communicating and planning *for* disaster, best practiced when you're not mid-crisis. Here's a planning tool to use at the level of your own household.

Setting the Table

You'll notice that this set of questions is longer and more specific than some others. That's because it's helpful, as far as possible, to have things spelled out, including needs and desires that we're not used to spelling out—even with the people we love most and know best.

QUESTIONS

Who *that you know* might come to you for shelter if climate change drives them from their home?

What kind of room for them could you make? For how long?

What would you need them to know about living with you, and what would you need to know from them, in order to live well together for that length of time?

How would you help them find their next landing place *before* that time ended, or if you couldn't live together as long as you expected?

How will you handle disagreements?

How will you let them know they are welcome—what care will you offer?

What would you want to cook for them the first night, if you had the option?

PRACTICE

Ask and answer the aforementioned questions with all of the humans you live with, including children and elders. You'll probably disagree at first, so build a consensus—you may need to sit down together with these questions a few times. You can use the answers to guide a later conversation with people who might come to stay, or even people whom *you* might need to ask for shelter. Each time, cook and eat together the food you might make for your guests.

If you're part of a group that does these questions and practices together outside your household, come together

> for a shared meal, and troubleshoot both recipes and strategies. If some people need to participate remotely for reasons of disability, illness, or immune safety, drop some food off for them, or cover their grocery list for the dish they want to make. Pooling money for costs can also make this more inclusive for people who are short on funds. Help each other figure out how to cook in bulk if you need to feed more people and how to adapt dishes for allergies, if certain ingredients aren't available, if you have to cook on a grill or a campfire instead of a stove...

The other thing I've learned from housing people in need, as well as from listening to other people who've done it and people who've needed it, is that you don't want to have only one option or be anybody's only option. Look at S and her family from the beginning of the chapter, stuck in the same space for longer than was good for any of them. People who are coming from trauma and disaster, short or long, are reeling from rupture and danger and loss: they're not always bringing their best to the encounter. And people whose homes are relatively stable need to clearly understand the power difference between them and people whose home is in flux. If you can kick someone out, and they can't kick you out, you need to take extra steps to make sure you're on an equal footing in other ways, and one of those ways is making as sure as possible that they'll have somewhere else to go if living together becomes intolerable.

Sources of conflict might include trauma responses, shifting boundaries, miscommunication, and competing needs, as well as strains on money, food, and space. You can't count on this change leaving your relationship entirely the same. But if you're housing people you care about, you have a reason to work it out, and a reason to help them find alternatives if you can't. And if you end up being the one seeking shelter, having answered those questions and having had those discussions will still leave you better prepared to live with people in less-than-ideal or unexpected circumstances.

By knowing what we can offer and accept freely, we're less likely to see and respond to other people as threats to something we're trying to *preserve*: time, space, food, value or worth, power, or just the way things have been. Climate change threatens to make everything familiar unfamiliar. So another way to meet it with purpose and care is to make yourself at home even in altered, interrupted, and continually changing circumstances.

BECOMING AN ARCHITECT OF HOME ANYWHERE

Home is where we do the things that are familiar, and doing those things—ordinary and also holy—can make us feel more at home wherever we are. My student Stina spoke about remembering a friend after they'd both left their city: "We used to walk out across the Brooklyn Bridge together, and back over the Manhattan Bridge together. And now I can still feel her in my feet when my feet get tired." Community development expert Angela Blanchard relayed to me the story of Patricia Jones, who held a funeral at the levee break for everyone who was washed away and never found after Hurricane Katrina. Cindy Quezada, whose family is from El Salvador, worked at one point for the Central Valley Immigrant Integration Collaborative in California. She told me about getting invited to festivals held by the farmworkers she knew, who are mostly from Michoacan, Oaxaca, and Guerrero: dressing up a figure of the Christ child in new clothes for Día de la Candelaria, parading with a statue of the Virgin, invoking their various Indigenous traditions through costume and dancing. "People create their place where they are," she said.

Ritual—formal or habitual, personal, or shared—is one way that we store meaning in our actions. All something needs to be a ritual is for you to do it multiple times, more or less the same way each time, with intention and attention. You probably have multiple rituals in your life, and even more small common actions that you could fill with meaning. Here's an exercise, with a choice of two practices, for ritually bringing your connection with your home (whatever that is and means to you) into focus.

Strengthening Muscle Memory

This exercise lets you carry your sense of home with you in a way that you choose. Even if you spend the rest of your life in the place where you live now, the practices can refresh that place and its connections and relations for you, making them more nourishing to you. And of course if you have practices of ritual and reconnection that you already use, you can draw on those instead, or weave them in with these.

QUESTIONS

Which parts of your sense of self are portable in your own body? (Songs, recipes, dance moves . . .)

How do you move in the places you love, with the people you love?

PRACTICE

Walk, or roll, through the place where you live, on a route you often take. Be deliberate about getting the memory of the place into the motions of your hands, head, back, feet—whatever parts of your body are involved in propelling you or in feeling the motion—and into your heartbeat and breathing. Do this for several days in a row, or as close together as you can manage. Finally, come up with a motion, gesture, or dance move that condenses that route or memory for you, and brings it back into body and mind. Do this motion whenever you want to feel your home.

Or

PRACTICE

Choose an activity involving motion that connects you with the living beings who feel like home to you: preparing food or formula, petting an animal, propping a window open, watering a plant, giving an injection... Come up with a motion, gesture, or dance move that sums up that activity for you, and brings it back into body and mind—say, the motion of watering the plant, even if the plant isn't there, that holds that same feeling of grace and care. Do this move whenever you want to situate yourself among the relationships that make up your home.

If you're doing either or both of these as part of a group, you can bring each other along on the process of developing the gestures, practice the gestures together—or create a shared dance routine that involves all of them.

My family is Jewish in a partly assimilated way: we *feel* Jewish, but religious practice isn't a major part of our daily lives. At our Passover Seder—the ceremonial storytelling, prayers, and meal—the "lamb shank" on the Seder plate is painted cardboard, God's pronouns are gender-neutral except when they're not, and the story is missing some pieces. But one of the lines we've kept is, "Know then that your people shall be strangers in a land not their own." The people in our family who survived the Nazi occupation of France, where they were immigrants, were sheltered by neighbors and nuns. Some of them lived a long time, and lived fully, after the war.

The Passover story doesn't relate who was living in the Promised Land when the Jews got there. Year after year, that information is missing. Many peoples, from the children of kidnapped and enslaved Africans to settlers and colonizers on multiple continents, have used the Passover story for strength. Some Jewish people have read it as a reason to occupy the land of Palestine, creating an armed border that kills people both quickly and slowly. Like any powerful story, it matters how its power's being used. In the story, God parts the ocean for the people seeking freedom. In this world, we must part the ocean for each other.

CHAPTER 3

MANAGED BURN

FROM ALIENATION TO INTIMACY AND STEWARDSHIP

YOU ARE HERE

This chapter invites you to do these things:

- Pay careful, receptive attention to the plants, trees, bugs, fungi, animals, and humans where you live.
- Learn more about the ways they care for you, and the ways that you can care for them—and how doing so will help all of you weather the changing climate, together.
- Learn more about the history of land, water, and people where you live, and how healing that history can bring about a better, more just, and more livable future.

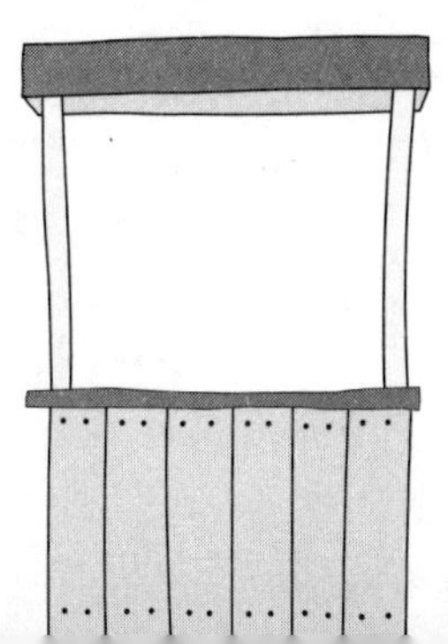

RECOGNIZING YOURSELF AS PART OF AN ECOSYSTEM

The Miwkoʔ people have knowledge of sea level fluctuations going back through geologic time. That's thousands of years of many minds adjusting, observing, forgetting, restoring, recalibrating, forgetting again, reminding, filling in. "We're river people and tidal estuary people," Miwkoʔ scientist Don Hankins told me, "so we move with the tides, and our creation story tells us about the place where we exist, that even at climate change maximum we're within our story scheme. We're still able to be part of the tides and the rivers, within the place that our story tells us about." For most of their endurance as a people the Miwkoʔ have ranged and circulated toward and away from the north-central coast of what settlers call California, according to their needs and the seasons and conditions. "And that allows us to have the flexibility to continue to persist as a culture and society within that place."

When Don's daughters were learning about weather in third grade or so, he encouraged them to read the clouds as Miwkoʔ have done since long ago. "You see those cirrus clouds forming?" he'd ask them in the fall. "What does that tell us about what's going on?" He described these shifts in the clouds' shape and texture as symbols "that change is gonna be coming—in the old days that would maybe be an indicator for going out and setting the landscape on fire" in purposeful and managed burns, timed to wind and weather, that limited more damaging fires and made way for new life.

Don's knowledge is both generational and cultural: the interpretations he heard from his family and community members, and the ones he passes to his children, are in constant dialogue with what he feels and sees. Much of that knowledge is also in dialogue with formal scientific work, and he shares it with students: he now teaches courses in fire, water, and land stewardship at CSU Chico in Butte County, well north of where he grew up on the San Francisco Peninsula. "Even though I'm not from here, I know enough of the things to pay attention to," he said. "Living in the mountains, the wind patterns are different than what I'm used to in the valley, but I also know to pay attention to learn those things." His observations are careful and thorough, day by day and hour by hour: "Generally in the afternoons, the wind is gonna pick up, and the winds are still maybe coming from down in the valley, coming upslope, but by a certain point in the afternoon it's gonna shift."

Offering and sustaining that kind of attention to the living systems you're part of, as climate change changes them and you with them, can eventually help you work together for your shared well-being. And the attention itself is healing—if not for the landscape yet, at least for you. You can begin here.

Placing Yourself

Ideally, this exercise will put you into an alert, relaxed state that also supports clearer thinking and decision-making. Every leaf and every bug, every rivulet and every sidewalk crack, is full of variety and activity and pleasure that you can share. Even if this is something you already do, I encourage you to do it more, and to let it impress and delight you.

QUESTIONS

What parts of the place where you live do you regularly notice—how they stay and how they change? (The weather, the temperature, plant life, insect activity...)

What parts do you notice only when someone else points them out?

PRACTICE

Over the next week, be more deliberate and conscious about the attentions you already give. Keep a notebook if you like, or notes in your phone. The following week, add one element or part to pay attention to: if you already notice insects, add clouds.

If you're doing this as part of a group, you can text each other when you notice something particular. You can also make a point of paying attention to each other's things: if you don't usually notice fungi or the flow of surface water, but someone else does, now you both do.

When I visit my parents in the house where I grew up, my mother and I walk around the yard together, noting growth and death. In the winter, through the trees, you can see the lake of Lakeville, which the Mohawk people named Wononscopomuc. We stop to peer at the top rail of our neighbors' fence, where the lichen sometimes called British Soldiers is growing. "Noticing things like those little red dots on the top of that moss, or lichen, whatever it is, is just a thrill to me," she said. "And I think that having someone—meaning you, in this case—who was receptive to that, encouraged me to continue, to increase it." We've been walking in the woods and fields together since I was very small. Even with her hearing aids, a lot of bird sounds escape her now, but she still likes me and my sisters to tell her about the ones that reach us.

In Flatbush, Brooklyn—her concrete-heavy, lifelong home—poet, playwright, and grief expert Diane Exavier reminded me that "living on the land is also living with people," our bodily needs and demands, our lives and deaths, what we get from and give to each other, how we get in each other's way. All the work you did in the previous chapters, the connections you wove and the ground you prepared with other people, is also ecological work. Slowly, as you learn where the water pools when it rains or how long your neighbor's son goes between visits, the way you're surrounded by and within life comes into focus.

Within climate change, the truth that all of us are parts of nature affecting one another matters profoundly. It reminds us that *all* landscapes, *all* ecosystems, are worth tending to and caring for—not just those that are "pristine" or lush or used recreationally by wealthy people. But if you live in the United States and are shaped by its dominant culture, you might hold the notion that "the environment" is *something else*, something you stand apart from. The day after talking with Diane, in another part of Brooklyn, I overheard two kids walking around the corner under the raised train tracks and saying, "There's no nature around here!" I knew what they meant, and the lack of greenery and other life does take a toll on human minds and bodies. Yet both of those kids were nature: the one saying, the one hearing. And so was everything around them.

LETTING GO OF HUMAN SUPREMACY

When Jolon Indian journalist Deb Krol writes about Indigenous life, politics, and land, she doesn't separate those three categories. Born and raised on Jolon Salinan land on the central coast of California, Deb lives in Phoenix now and talked with me while we sat in plastic Adirondack chairs—the ones that tip you way back—outside the hotel room where she was staying, in Narragansett territory, for a climate journalism conference. The chairs are named after the mountains; the name of the mountains comes from the Mohawk language and probably means porcupine.

After the conference, Deb was bound for the north coast in California. There, she told me, a group of people from several Indigenous nations and tribes were acting together in defiance of settler conservation policies, made with disregard for the relationship between people and land. People who wanted to maintain and restore that relationship were meeting "to continue gathering for subsistence and to gather traditional foods and to make regalia, and to restore their Indigenous food sources, like the push to restore salmon runs. Because up in northern California, the salmon and the people are one. And without salmon, your people are not going to survive."

Acting like the elements of a living system don't depend on each other means you have to deliberately ignore a lot of evidence. But many of our social, political, and economic systems are set up to do just that, and convince us to do it, too, as an excuse for treating living systems and the people in them like *resources* and extracting their value. Human supremacy, and white supremacy in particular, is a myth that helps to justify that extraction. When Europeans colonized this continent, that was *their* creation story, fed by Christianity and monarchy and conquest, and they brought it with them.

I grew up tangled in that story, too, looking at classroom posters of tree frogs and zebras set within only the tiniest glimpse of their homes. Even as my mom and I were examining deer poop and lichen and mole tracks together behind our house, breathing oxygen exhaled by trees and grasses, I learned to think of "the environment" as something I was at most *in* but never *part of*, something I could *save* without saving myself. Here's an exercise to let that go, and to make visible the webs of relation that you're in with everyone you know and every living creature in the world.

Knowing Your Neighbors

This exercise is so much more bountiful with a group: each participant brings in the relationships they notice the most, and you get to see the complexity of the connections we live among much more clearly. It's a great one to do with kids, too.

QUESTIONS

Which nonhuman presences—plants, bugs, animals, water, wind—say "home" to you?

How do you feel their presence, whether you're together or apart?

PRACTICE

On big paper, a chalkboard/whiteboard, or a wall or sidewalk, map out a web of relationships that you're already part of: start with yourself, write the living beings you're connected to, and draw lines between. Be sure to include all kinds of beings and relationships: the plants and animals you eat and the people who raise them, the trees or grasses that keep the ground you live on from washing away, and the bugs or winds that pollinate them. Look up what you don't know, or put question marks if you aren't sure.

If you're doing this with a group of people, make your map together, or combine your maps at the end. When you're happy with the web you've made, use chalk to mark it on an outside wall or sidewalk if you did it indoors to start with. Label it whatever you think will help passersby understand it.

Naming and knowing these webs of relation allows us to notice what they offer us, and what we might offer them. Ignoring these webs, on the other hand, lets us think that we can take from "our environment" without messing up our own lives, interact just with the parts we choose, and control our experiences with no consequences. Pouring insect killer on one lawn isn't the same as saturating acres of crops, and the people who harvest them, with pesticides. Carelessly letting off fireworks in fire season isn't the same as a multiyear policy of fire suppression that ignores Indigenous knowledge and observation of changing conditions. But they have the same cultural root in thinking that humans don't have to understand or care about the living systems we're part of and how our actions change them.

Climate change is happening *because* we're part of the world, and what we do affects it. The illusion of independence from and control over "our environments" leads to inequality, suffering, and unnecessary death. It also distorts our thinking and is part of the isolation that we already know is both dangerous and untrue. Releasing that illusion feels risky because some of us benefit from it in the short term—less expensive food, convenience or efficiency in our work—and sometimes, as with air conditioning on a brutally hot day, we have to make trade-offs for our survival. But hanging on to the illusion hurts us, and hurts the lives around us. It deadens us to relationships that already exist, that we could pour more love and care into, and that could sustain us.

PLACING YOURSELF

Bringing life back into your connections with the living world might be painful, like when blood flows back into your foot after it's been asleep—or like opening your hand to receive something new, after clutching hard what you were holding on to before. Once you've aligned yourself with the living world, you care about it: really seeing a particular kind of tree, you may also soon see that the ones near you are sick and dying, and grieve for them in a way that you wouldn't have before. This is the risk we take in knowing and needing each other. And after the risk, there's the maintenance: learning, acting on what we've learned, learning again as new facts come forth.

In her tribe's philosophy and ethics, Deb Krol told me, humans are part of the Earth, with a different kind of consciousness than anyone else here, and we were put here by the Creator to use our consciousness for the well-being of the totality of the Earth. That might sound at first like the human supremacy I just asked you to let go of—but *different* doesn't mean *better* or *superior*. It means *particular*. It means identifying what you're good at, and good for, in your dealings with other lives. Our roles aren't interchangeable. If I was tasked with pollinating a potato flower, say, that flower would be waiting a long time.

Deb described what I'd heard from Don Hankins as well, connecting the prohibition of managed burns (which people indigenous to the West Coast used to do regularly) to the increase in fires that are also disasters. When the overgrowth isn't purposefully burned off, "a deer can't get through there, there's no food for it. The woodpeckers can't eat, the squirrels can't eat, and they go somewhere else. And they're not performing their role of redistributing seeds and repopulating beneficial plants. And then stuff gets overgrown. And a drought comes along and it gets dried out, and it becomes tinder, and the big wildfires come through. Well, in my point of view as an Indigenous woman, a lot of this is because the Earth knows it's sick. It knows it's not working right. And it's trying to fix itself. But because it doesn't have that consciousness that humans have, who were put here to take care of things, it doesn't realize that when it allows these fires to come through that it's gonna clean things up but everything in its path is gonna get destroyed."

In this view, care—attention, intimacy, then action—is a human obligation, our role in life, within all of life. Community care, the kind that my neighbors and I exchanged at the Climate Anxiety Counseling booth, extends to our human and more-than-human neighbors, and builds a connection that actively strengthens us. Diane Exavier describes her frustration with people who live “as if they were the Lone Ranger. They’re looking at people who have only survived collectively and going, ‘What is this magical thing you have going on?’” The sister who feeds her cat when she’s away teaching, the priest that her mother cooked for during seminary who then signed off on her godson’s baptism: these small, ordinary motions and relations add up. Intimacy with the place where you live is the step between attention and action: getting to know both what’s happening there and what it means.

There’s also the view that intimacy is the basis for a living system of *co-governance*, where each form of life in a place contributes to the overall order of life in that place and has knowledge to share with the others. Scholar and professor of law John Borrows, a member of the Chippewa of the Nawash First Nation, has written that across many widely varying Indigenous systems of law, he sees a common thread: “a process of identifying what we see as necessary in the natural world to be able to take our cues, to live well and helpfully with one another.” If law involves enacting the responsibility we bear to one another, more-than-human beings—lives that aren’t human, but aren’t *less than* human—can be humans’ teachers in this. Replanting after a fire needs to be responsive to the climate shifts that have already taken place and are predicted: plant profiles and relationships must change because growing conditions have changed, and will change further. People who align their efforts with land and water need to learn a lot: how to recognize the overall health of the meadow or forest or chaparral or desert, who lives there and how many of them, how to tell when it’s well and when it’s sick, what kinds of care will matter. You don’t need to keep that all in your head. You can learn it from them.

Don Hankins reflects on the changes he and generations of his family have seen in their lifetimes along the Bay and region. "I pay attention to those shifts," he said. "When certain birds are coming in to nest, and when I see certain insects. You kinda take the long-term view—Okay, well, last year, this species bloomed a month earlier than it normally did, and that's kind of been the trend of the last ten years. Well, maybe that's a signifier of that shift."

Our reasons for learning what to pay attention to are many and complex. The need to survive grabs us: recognizing that we're about to have a seizure or that someone's temper is about to snap, operating a forklift or a deep-fryer without injury, remembering where to dig clams or buy the cheapest instant noodles. We also pay attention when someone we respect tells us what to look for, and how, and why. Even if we don't immediately see how we fit into the well-being of the place where we live, we start looking for where we fit in, because we believe them.

In 2014, when I set up the counseling booth at a county fair in southern Rhode Island, and news media weren't yet reporting wildfire after famine after flood, a youngish white man told me that he knew climate change was real because "people older than me [are] telling me that things are blooming earlier. Ponds you used to be able to skate on all winter melt after a couple of days. I'm not old enough to have that multigenerational perspective." He'd listened as well as seen, and used that knowledge to increase his connection both with his surroundings and with the elders themselves.

This sharing of knowledge can happen even when it seems like we don't share anything else. In 2017, a guy who told me that he "would be Trump if [he] had the money" also told me about a man he'd spoken with on Rhode Island's Narragansett Beach, named after the tribe whose land it is: "He's Native American and he's lived here his whole life, he's seventy-two years old. And he was telling me that all the way on the right side of the beach, past Chair 1, that used to be sunbathing territory. Now it's one and a half feet deep at high tide. It hits the seawall. Even at high tide there used to be fifty feet of beach there."

In exchanges like this, what you're sharing—maybe across time, maybe across cultures—is both knowledge about the past and the present, and an affirmation of your environmental perceptions. That conversation gave the guy at the booth another way to understand his world and to know that he wasn't nuts. And talking to me about it let him hear how it sounded to change what he knew, and make it a little more real.

None of that, in itself, removes inequalities or repairs violent histories. I don't have any expectations that the guy at the booth, who was white and in his sixties, will now respect the sovereignty or fight for the well-being of the Indigenous man he spoke with—that he will change, or be changed. But *you* can, if you allow yourself to listen this way to the other living beings who live where you live, and who take part—even unequally—in the risks and riches of that place.

Connecting Through Generations

This kind of conversation is an opportunity for generosity and sharing, and it can add to your knowledge and connection even if you're also an older person: you may notice different things based on the lives you've had. It can also come with tension, grief, and frustration that are rooted in history. Notice your inner reactions as you listen, and don't feel obligated to outwardly respond until you know what you want your response to be.

QUESTIONS

Who's an elder in your life that could share with you their remembrance of the rest of the living world and the changes they've witnessed?

What have they already been saying that you could listen to?

What could you ask them?

PRACTICE

Ask an older person who lives where you live now what they remember about how the place used to be, and what changes they have noticed. If talking with them makes you feel closer to them, notice that. If it makes you feel defensive, resentful, tired, or sad, notice that, too.

If you're doing this as part of a group, there are a few directions you could take it, from pooling your resources to do something that the people who are telling you these stories would enjoy, all the way up to becoming part of each other's evacuation plan for a climate disaster. The point is that connecting with (other) elders *is* connecting with the ecosystem you're part of: you are all there together.

That may mean being together in pain. People who have felt helpless to preserve the aspects of their lives that they needed or liked may tell their stories with anger or bitterness; some ugly context may emerge, and they may find themselves facing or trying to avoid shame. Severing generational knowledge and human-land relations has also been an active tool of genocide on this continent—murdering people and driving them from their ancestral lands, separating families, slaughtering bison, and damming rivers. I learned through our conversation that Deb Krol and Don Hankins know each other, and that he had described to her how the salmon rivers worked to sustain the ecosystem when they were flowing freely. "Those are the types of things that are going to save Indigenous peoples," she said. "Knowing how the Earth works, knowing how we work with the Earth . . . that knowledge, that pulse, that we know how things are supposed to work, and our push to restore them back to where they should be, that will save our people."

Deb sees some possibility in sharing this knowledge with non-Indigenous people, but she also has some doubts. Leaders and educators she respects—Ron Goode of the North Fork Mono Tribe, Mona Tucker of the yak tit^{y}u tit^{y}u yak tiłhini Northern Chumash Tribe, and Kyle Powys Whyte of the Citizen Potawatomi Nation among them—have all said to her that the Indigenous work of restoration is not just for their own peoples and the lands and waters with which they evolved, but for the descendants of colonizers and invaders as well. "We Indian people," she said, half-quoting those conversations, "we're already in a post-apocalyptic world. Our apocalypse has happened. And now we're picking up the pieces and trying to rebuild our civilizations. And so, because we already have this knowledge, we can be of use to the rest of the world when their apocalypse happens."

"I've heard that too," I said. "How do you feel about it?"

"I'm understandably a little bitter," she said. The long violence of colonizers, their descendants, their attitudes, and their practices have attacked Deb and her family and her people at every level, from state law to medical negligence to individual contempt. Deb's great-grandmother lived and died refusing to admit that she was Jolon because of what being Indian meant in a state with genocidal policies on the books. So it's not just that they had different lives than other residents on the land—it's that some of those residents were actively working to make their lives unlivable. Colonization and genocide are why my parents' town is called Lakeville and why its soil is full of iron slag. It's part of why my white and Jewish ancestors came to this continent—they didn't ask for that initial violence or take part in it, but the land was available to them because of it. All of us live in the world shaped by colonists' decisions and the decisions of their descendants. Making different decisions, which can help us live better within climate change, also means setting right some of those wrongs.

The UN Intergovernmental Panel on Climate Change, which published the reports that got a lot of people frightened about climate change who weren't frightened before, has been very clear that the lands that are healthiest, liveliest, and most sustaining across the world are managed and cared for by people indigenous to those lands. At the community level, this is a reason why negotiating land return or Land Back is both just and wise. Like the resumption of Tuluwat Island, more recently held by the city of Eureka, by the Wiyot tribe. Like the return of sacred lands along the Rappahannock River to the tribe that bears that name. Like the renewed stewardship of Tc'ih-Léh-Dûñ, a redwood ecosystem, by a council of ten local tribes. Like the partnership of nonprofit collective Movement Generation and the Sogorea Te' land trust, returning forty-three acres of land in the Bay Area to Indigenous care.

Complicated? Almost definitely, as everyone involved navigates loyalties, priorities, and differences in power. Costly? In some ways—not just in money, and not just to the people who are relinquishing money or land, or contributing time and work to caring for the land with Indigenous guidance. But there is nothing impossible about it. And doing it might make other things—things we all need, whoever our ancestors were—more possible. Land Back can make our actions to care for the living world more purposeful, because they're better informed and no longer working against one another; it can nourish a root system of alternatives to the structures that injure us. Here's an exercise to help you explore it, whether or not you're indigenous to the place where you live now.

Preparing the Land for Restoration

How people report feeling when they do this exercise depends, not surprisingly, on their position in history or in the present: if they're Indigenous and on their own land, or if they've departed or been displaced; if they're the descendants of settlers or colonists; if they're more recent immigrants who stepped out of one web of history and ecology into another. Whatever your emotions and sensations, feel them in the present moment, without judging yourself, and without jumping to act on them. This exercise is also an opportunity to revisit your assumptions about where you live: What have those assumptions been? How have they changed? How *could* they change—how might you let them or invite them to *be changed*—through what you observe and are told?

QUESTIONS

Which peoples are indigenous to the land where you live now?

Are they, or you, presently in negotiations to resume stewardship of that land?

PRACTICE

If there's an ongoing Land Back effort in your area, and you're not already involved, reach out to see how you might be able to support or advance it. If there isn't one, there may be an option for the descendants of settlers to formally pay rent to local tribes or to make regular donations. If you are the title-holder for land or property, you can also research deals that private landholders have made to return land to the people who evolved in step with it, while still living or working there with the guidance of the tribe or nation. Indigenous readers living on others' ancestral lands can explore the potential for change where your relatives are from *or* where you're living now, or both. Keep a record of what you learn, especially as it applies to the place where you live.

If you're doing this as part of a group, you can hold a research gathering or research separately and pool your findings. You can also talk with each other about how the process is affecting you emotionally, keeping in mind that your different histories and experiences as residents of this land will affect how you feel, and how others respond to what you say.

A full treatment of the movement for Land Back is larger than this book will hold, so for the time being, the research and the process of feeling what it brings up *are* the practice. If you want to be part of local Land Back efforts but don't have much experience being organized by others, there are guidelines in Chapter 6 for matching yourself up with an existing group. One thing I have learned through my years of listening is that what we offer voluntarily is different, different in *nature*, from what is taken from us or what we cling to no matter what it costs. That applies to our senses of self, our dreams for the future, our care, our space, our time. It applies to the land, too.

BEING WELCOME TO THE LAND

There's a locust tree in my parents' backyard, barely level with the top of their house when they moved in, now towering and spreading. Its huge quantities of tiny leaves, yellow when they fall, copper when they dry, make my mother's favorite garden mulch. "It's an ongoing relationship that I've always loved," she told me. "I'm not attributing [the tree] to have any sort of consciousness about me, but I have a consciousness about it. And just in case, I greet it, I pat it. I lay my hand on it, and I don't know if I feel any vibrations or not, but just in case."

In addition to his work on his own people's land and other tribal or once-tribal lands on this continent, Don Hankins has also worked for a long time on Kaanju Ngaachi lands in Australia, together with the Kuuku I'yu people whose homeland it is, to study fire and biodiversity there. Something that happened there changed the way he teaches his CSU Chico students. For many years, he didn't talk in the classroom about what he calls the metaphysical aspects of nature—the sources of meaning that academic science can't measure—"for fear of people not valuing my culturally informed approach as a scientist." But in 2015, when he was out on the land with a group of students, four rare birds of cultural significance showed themselves. Their presence was a mark of approval to the community, Don's team, and their work in revitalizing fire in this place.

"Culturally, I'm used to that happening," Don said. "You just take for granted that certain species will show up to kind of give approval to the work that's going on," if the work is in line with the well-being of the place and the relationships that form it. "But the students didn't have any awareness of that. And things were happening that they didn't have the preparation for."

Don began including some of that preparation in his courses, so that students will have an additional way of knowing whether their work is good for the land. In courses on pyrogeography, conservation, restoration and stewardship, and water resources, he teaches that a particular species, in a particular place, can be a messenger. "People may shut down, or they may get excited about it," he said. "And I tell students, regardless of what their belief system is, or their way of understanding the world, that if they're open-minded about it, the world will then communicate to them in these different ways." Here's an exercise to offer this kind of informed, directed care and co-governance, rooted in what you as a human *can* know and do, for the part of the world that cares for you.

Making an Offering

This exercise will continue to cultivate your intimate knowledge of the place where you live, where you fit, what your limits are, and how others are carrying their parts of the load. It shifts your sense of what really is nonnegotiable and what can change. And it shows us how to align ourselves with, to *be organized by*, the cycles of life and death and growth and rot, how the soil holds water, how the air passes through—instead of fighting them for control. Not having to do or be everything is not just a limitation, but a gift and a blessing.

QUESTIONS

What would you like to say to the place where you live, and the living beings there?

What could this place receive from you that it would like to have, and that would build your connection? (Watering, compost, protection from development, trash pickup...)

How would you know if what you offered had been "heard"? (More light, different smells, more plant growth or insect activity, other humans joining you...)

PRACTICE

Based on your answers to those questions—which can be informed by the other exercises in this chapter—create a small ritual of care for the place where you live, or a part of it. It should feel meaningful and have a good effect, and be something that you can make part of your routine without much strain—it's better to have something small that you will do than something large that you won't do. As you continue to do it, notice both how you feel and how the place seems.

If you're doing this as part of a group, you can create a single ritual together, or attend each other's rituals.

"'I need to fix the planet,'" Diane Exavier said a little scornfully. "You don't see the tree looking at the river and being like, 'I'm gonna river for you too.' Imagine showing up to the planet and being like, 'I'm gonna save you.' You need to be addressing your brother. The planet would be like, 'Thanks for doing *that*.'"

In October of 2018, I went with my friend May Babcock, who has a lot of local plant knowledge, and Fish and Wildlife Service employee Ben Gaspar to collect salt grass seeds at Goddard Memorial State Park in East Greenwich, Rhode Island. We were taking the seeds from one marsh in order to restore another, after waiting for enough seeds to drop to maintain this patch for next year.

Coastal peoples know, and more and more settlers are catching on, that wetlands and marshes make the meeting of land and water more flexible, both elastic and lasting. Shorelines that include them handle flooding and storms with less damage. Over the winter, Ben sent the seeds to a greenhouse, whose workers soaked them in cold saltwater and got them started in little cups of soil. And in the spring, May and Ben and I joined other volunteers on the banks of the Narrow River to plant the salt grass seedlings. About halfway through, I started drawing a circle in the sand around each planted seedling and saying words to it, a prayer for its good growing that I invented on the spot and don't remember now.

Another friend with a lot of plant knowledge, who grew up at the mouth of the Narrow River, told me that previous years' salt grass plantings hadn't taken hold. So I went back to check. The shore mud was stinking in the July heat, with a sheen I've been told is bacteria, not an oil slick. And the grasses were happening. Rooting into the sand and soil, collaborating with microbes and insects, and—as I looked closer—sending out runners, which I could tell because there was a line of mini-grasses, living and unmistakable, coming up between two of the plants we'd set.

That is how grass likes to spread: by secret underground pathway. Seeing it living where I'd planted it felt amazing. But time will show whether gathering the seeds and setting the plants was in itself good, not just for me but for the grass, the river, the bay, the world. The purpose of gathering the seeds was to restore the grass. The purpose of restoring the grass is to enrich the remaining land, water, and smaller living beings. No restoration means no enrichment; no silting up of nutrients, no slowing down to a livable pace the shifting of water and Earth; no buffer against the violent interactions of water and air that come when a storm comes.

Will the grass hang on to help its surroundings survive the hurricane, the drought, the wintertime, the hotter summer? Climate change means that what I or you try to be part of may not endure. But however long these relationships last, they grow, and they give back to us. I would rather live in this mutual, generous way for an hour if an hour is all I have. If the world and the part of it where I live have fifty years to share with us, or thirty, or twenty, I want to live in relation with the rest of it until I turn back into dirt and feed into it in another way.

Think about the great, slow, rolling changes that Don Hankins assured me that the Miwkoʔ people lived through, how they ranged and readjusted, the elements of their culture that must have emerged and receded in response to their conditions. They suffered and lost and grieved throughout those thousands of years of change—not as much as they did through the catastrophe of settler domination, but still some. The point is not that we will be okay. Nobody knows that. They didn't know, either. The point is that they—your ancestors if you're Miwkoʔ and reading this—did their best to be part of their world. It makes sense for their descendants *and* the rest of us, even those who fear we're counting our survival in years rather than in decades or centuries, to be alert and take care while we're here.

CHAPTER 4

A CONFUSING BLESSING

FROM GRIEF TO POSSIBILITY

YOU ARE HERE

This chapter invites you to do these things:

- Enrich your relationships by being truthful about climate change and our place within it.
- Embrace futures with different hardships, and different satisfactions, than the present that you know.
- Let go of, and grieve for, visions of the future that are not compatible with climate realities.
- Follow your grief back toward the love that prompted it, and use that love as a guide to purposeful action.

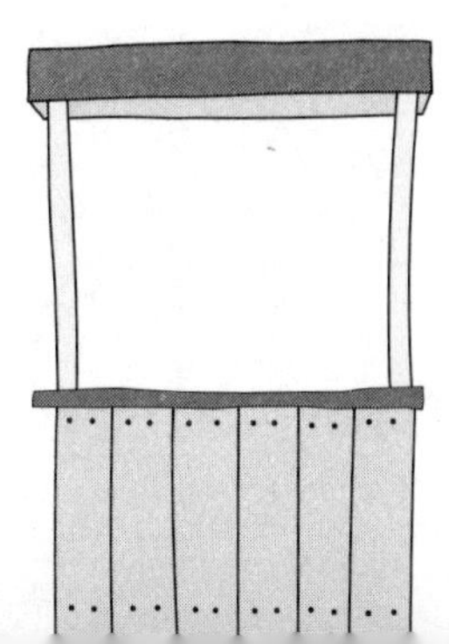

THE FRAGILE FUTURE

"I'm in the midst of grappling with what we have to say goodbye to," said the stranger's voice on the phone. Dezaraye Bagalayos and I had been talking for all of five minutes, and we were getting right into our climate grief. "I feel like that's pretty much a collective response, pretty reflective of the world right now. We're grappling with what we're going to be losing. I think that we don't even really know where it begins—we can't even think about where it ends."

Dez works for water justice in California's Central Valley, Yokuts land. When we first talked in January of 2019, our shared preoccupations with climate collapse converged in a way that was a relief to both of us. If you absorb the ways that climate change is changing our world—because you're forced to or because you choose to—you're bound to feel some degree of *ecological grief*: the ache of the lives and relationships already gone from the world, and the anticipation of further losses. You're also bound to feel the more intimate pressure climate change exerts on the life you were preparing for yourself and, if you have or want them, for your kids. When someone else tells you they're feeling those pressures, those losses, it's a confusing blessing: someone else is as sad as you are.

A few months later, Dez and I sat in her kitchen drinking tea out of huge mugs and talked more, talked for hours, as though we would never get another chance. I got to her house in Visalia, California, as the spring dark was falling, the air soft and clear; she told me it was one of the last weeks of the year when they could count on good air in the Central Valley. For my visit, she'd set up a tour of the Tulare Lake basin with one of her mentors, ecologist

Rob Hansen, and other people who could tell me about how the changing climate was affecting water, fire, land, and people in that part of California. I hadn't thought about the air at all; Dez had to breathe it all the time.

"I have these conversations with myself, but not with another person," Dez said, about half an hour in. "I talk about it with my partner, but we're usually talking about it in reference to our daughter, like, 'What in the hell did I do to this person!' Like, 'You're the best thing that ever happened to me, but what were we *thinking*?'"

Dez and her partner, Mike, share an Evangelical and apocalyptic Christian background that they've both broken away from. When she got pregnant in 2012, "we were tired of living under this doomsday reality. We were like, 'We're just gonna immerse ourselves in life.'" Their daughter Eleanor was born in 2013. Two years later, the United Nations' latest report on climate change highlighted how bad things were getting and how much worse they could get. And the following year, Donald Trump and his cabinet started undoing federal environmental protections—a process that continues under Joe Biden and the Trump-appointed Supreme Court. California was in the pit of a multiyear drought, and wildfires there and elsewhere were coming faster, lasting longer, killing more, and going further beyond what the land can handle.

People who come up to the Climate Anxiety Counseling booth, and people who speak and write about climate and the environment in more public ways, still sometimes talk about climate change in terms of their children or grandchildren, not themselves. When I started doing the booth, that was much more common. There was the yoga teacher who wondered whether her children would have children, the housing justice organizer whose stated worry was her teen's job prospects, the person holding their two-year-old son who introduced their climate anxiety with, "By the time he's in college..." But college might not exist when that kid is eighteen. Jobs might be wildly different. And even if she and all her kids make it, there are many reasons why the yoga teacher might not ever be a grandmother.

We know a lot about the effects that climate change is having and likely to have. How—and when—those will hit any given household, family, life plan, and whether it will be for the first or the fifth or the tenth time, is less predictable. On their kitchen wall, Dez and Mike had hung a list of things they wanted to do with the house, in bright marker on big paper. It included items like REPLACE FLOORING and PAINT BEDROOMS AND BATHROOMS, and also items like DROUGHT/FIREPROOF LANDSCAPING and WATER TANK/RAINWATER HARVEST.

Planning is easier if you presume the future's going to have a lot in common with the present, or that the recent disaster or shortage was an outlier: that you'll be able to buy paint and young trees at the big hardware store, or that your toilet will keep sending your waste into the sewer system. Because Dez teaches and advocates about water, land, and climate change, she's also planning another future, wanting to plant up her yard to meet the drought she's known and the fire she expects.

Our stories about our control over the future, like our hopes and fears, come from our families and circles, our histories of class and race and culture, the media we observe, the faiths we were raised in or later embraced, our ethical and political loyalties. And many of them no longer hold up. To be responsive to the realities of climate change, we need to recognize that it *is* our present and, in many cases, our past, as well as our big community future. And we may also need to let go of some of our personal futures—including the ones that are dear to our hearts.

Down the road from the small house where Dezaraye lives with Mike, Eleanor, and Mike's mother, Mary, giant pumps suck water from deep underground to soak the near-surface roots of tomato plants and pistachio trees. At the edges of these fields are billboards that say things like EVERYONE'S LIFE DEPENDS ON AGRICULTURE. Many of them are sponsored by lobbies against state restrictions on water use. Irrigation pipes whose tops were once level with the ground now stand waist high, shoulder high, head high: drawing out all that groundwater leaves the soil to fall in on itself. Dez's teenage students live in towns and work on farms in the Tulare Lake basin, which lost its namesake lake in 1898 after less than five decades of irrigated agriculture, mostly by European settlers and their descendants. Allensworth, a historic Black township, once sat on the lake's southeast shoreline; the water that comes out of faucets there now, when it comes out at all, is full of sulfur and other minerals, making it smelly and brown. During a recent structure fire in nearby Alpaugh, the use of just one hydrant seriously reduced water pressure throughout the town. I don't know what would happen if two Alpaugh houses caught on fire at the same time.

The people who own the big farms probably imagine passing them, or the money they get from them, on to their kids one day. In an effort to guarantee that, they're driving the pumps deeper and deeper, squeezing the land dry. At least some of the people working there are doing so with the intention that things will be better for their children than for them: more security, more opportunity. The people repairing the pumps, the people designing the billboards, the people writing the water policy all undertook their work assuming that it would have a desirable outcome for them and, in some cases, for those they love.

Climate change makes some of those outcomes impossible, and others unethical. The next exercise, which helps you practice letting those outcomes go, is a tough one for a lot of people, and may feel especially unfair if you've been trying to leave scarcity behind. Take extra care to answer the question honestly, and surround the exercise with grounding or steadying practices that you know work for you—"Drawing from the Well" on page 19, maybe.

Unplanning the Future

This exercise invites you to choose an end to an inner commitment: to the future you thought you could earn, the comforts you wish for, the standards of success you held. That ending process is painful, generous, and different from the one that's already driving people from their homes all across the globe. It's the difference between a theft and an offering.

QUESTION

What's one of your plans or dreams—for your career, family, finances, travel, or pleasures—that may not be right or realistic to pursue because of climate change?

PRACTICE

First, call on the support of someone or something that gives you strength: an elder or ancestor, a more-than-human being, a place, a past version of yourself. If you're able, try putting a hand on your breastbone or belly button while you're doing this, connecting you to yourself. Then vividly imagine *saying out loud*, "I'm not going to do [your answer to the aforementioned question]." Hear it with your mind's ear. Next, vividly imagine your life without this plan—both its goals and its outcomes. Notice your physical sensations and your emotions without trying to change them. It's also worth trying this exercise *again* a few days to a couple weeks later, to see if anything has shifted for you in the meantime.

If you're doing this practice as part of a group, you can share both the plans or dreams and the emotional reactions, or just the reactions. When other people share, be calm and kind.

One feeling that often comes to people through this exercise is resentment. Sometimes it's a resentment of timing: *My parents never gave this a thought. It's not fair that I have to!* Sometimes it's a resentment of inequality: *It's unethical to have a big house in a gated community? Say that to the people my mom cleans for!* Well, make it a real question: What door does it open if that's what I'm saying? What if I'm saying it to you?

Bodily ease and safe shelter ought to be equally available to everybody. Steps that make them more equally available are steps that every community and every government ought to take and that every person ought to demand, and making their availability more equal is something to prioritize as the climate becomes more chaotic. The stuff beyond that—the extras that can only come from someone being exploited or displaced, down the road or across the globe—maybe we can't afford that anymore.

If you're feeling that that's not fair, you're right. How hard do you want to work to keep it unfair? To tread water in a way that ends your nightmare, but only by ignoring—or creating—someone else's? The economic and social systems that we know best lay out that path for you; to transform them, you need to let go of, and mourn, the hope that as those systems are, they'll give you what you deserve. And if you've been living the dream up till now: it's always a good time to give up on something that hurts other people, to share what is here to be shared.

THE KID THING

Of course, it's more complicated than that. And there are desires we think of as personal that climate change complicates, too, because they too exist in relationship with the world. Here's the offering I made: I let go of the future where I have a kid. *I* let go, I decided, even though I'm married to the person I would have had a kid with, and even though we'd been trying, and even though I wanted one. My offering was not just parenting a child of my own (something that matters most intensely to me and my husband) but the time, energy, and attention that would otherwise go to that child (something that could matter to lots of living beings, if I use it right).

Having kids is not a luxury, and it's not a necessity. It's fine. How you treat your kid, how you raise your kid, the life you provide for your kid, the things you ignore for your kid—that's where right and wrong come in, because those things can be sustaining or detrimental to the kid or the world, or to you.

Many of the people you've met and will meet in these pages are parents. Of those, many say publicly that they do what they do for their children, to bring forth a world where all children can breathe and move freely. Being responsible for kids, in other words, is entirely compatible with climate action. I didn't think it would be compatible for me. Because climate change puts all the systems of our lives into uncertainty, it can tempt us to just try to secure the safety and comfort of the people we love. I could imagine myself falling into that temptation. I decided to steer clear.

Sometimes I resent people who decided to have children in full knowledge of this crisis; sometimes I'm grateful for the openness, the options, to spread my net of care more widely, although I haven't always taken full advantage of that openness in effort or in risks. By and large, I feel sad about it, but more sad that nothing has happened since that decision to make me change my mind.

People generally understand my grief for the life I would have had as a parent. So I sometimes use it as a gateway into talking about climate change—something that people feel along their veins, in the crook of their arm, where a baby's head would go. The heavy losses of climate change aren't equivalent to the loss of this specific, personal dream. But each can help us understand the other if we're vulnerable and honest about them.

In the summer of 2017, a young man in his twenties—white, tall, pale, in a button-down shirt—came to the booth in the park downtown. His opening: "Probably I have five years to live." When I asked, he said he didn't have scientific evidence for climate collapse within five years—he'd just picked something low. But his despair was clear. "The way I was raised was super hedonistic, just monstrously gaining things," he said. "I'm really privileged but I live with people who couldn't give less of a shit about the Earth. But I'm not gonna run into my parents' living room screaming, 'We all have to kill ourselves. Hey, mom, wanna go out and get some cigarettes and smoke until we die?'"

"It sounds like you're angry at your family," I said.

He looked shocked. "I love my family, they're great, I want to protect them. Imagine trying to love someone to your fullest ability in the shortest amount of time. You can do it by communicating, expressing your love, and you can even do it by silence. . . . I just don't want to have to do it. I don't want to have to do it all now. I can do it, but I realize how much love I was gonna have to give when I was older—I don't even think I'm capable of doing it right now—and I'm not gonna get to do that—"

He stopped because he was crying, and so was I. I came out from behind the booth and sat with him on the ground of Burnside Park, hard-packed city dirt studded with broken glass and cigarette butts. We held each other's hands and cried together.

I'll always be grateful to him for speaking about his loss in terms of the love he wanted to give. If the love is big, the grief will be, too, and it's hard to carry even the anticipation of it. If you've felt this weight of grief—or are willing to invite it in—here's an exercise to help you hold up under it.

Hosting Grief

Depending on your history and experience, this could be another tough exercise. If either the questions or the practice feel like they're going to reopen one of your wounds rather than treat it with reverence, back out of the exercise and do a grounding one—or try "Making an Offering" from page 68. And if you're doing this with other people and someone looks shaken, it's wise and kind to propose a pause or stop.

QUESTIONS

What is a loss you've undergone, or a loss you anticipate, because of climate change or the forces that drive it?

Where does that loss make itself felt? (Where in your body, your home, your landscape...)

PRACTICE

Arrange a corner or center of a room (or a square drawn with chalk on a sidewalk or driveway, or the front seats of a car) where a sharing of griefs could take place. You can add objects (a rock, a leaf, a lighter) and write words that hold meaning or memory, or make offerings of food or fire. If you can leave this altar set up, spend time there when you can—each day, each week. Again, you don't have to actively make yourself feel the loss while you are there.

If you're doing this as part of a group, everyone in the group can contribute to the altar, and join each other there. It can be the place where you answer the aforementioned questions together, discuss other things, or are present together without speaking. Make sure it's usable for everyone who wants to use it (chairs for those who need to sit, mobility aid accessible, no scented candles if someone's allergic).

This practice also allows the grieving and the not-yet-grieving to sit down together: our losses are of different kinds and depths, and yet the feeling of loss is something we can share. And because of that, when it's time, we can help each other up off the ground.

FINDING PURPOSE THROUGH GRIEF AND MOURNING

"The environment gives so much to me, I can't even pay it," said M. He was Latino, in his twenties, working for the RI Department of Fish and Wildlife. I'd set up the Climate Anxiety Counseling booth outside a Providence theater that was showing a movie about plastic in the oceans. "I love kids, and I wanna have kids but I'm so worried about my kids' future. What's the world gonna look like when they're my age? It's already bad now."

"How do you feel when you think about not having kids?" I asked.

"Empty," he said. "Without a purpose."

I asked what gave his life purpose now. Working with kids, he said, and "doing some stuff for the environment" in his job and his life. If he decided against having children, or found out he couldn't, he'd adopt: "I would help the kids that are here. I just really love giving back what I was given—people in my life have been so helpful, they've helped me find purpose and confidence."

The Work That Reconnects was created in the late 1970s by Joanna Macy and many colleagues, and has since evolved through the contributions of facilitators and practitioners around the world. It includes a series (they describe it as a Spiral) of internal and communal moves to prepare our spirits for connection and action in the face of ecological breakdown. The first, *coming from gratitude*, is meant to stimulate empathy and confidence, letting us take on the next moves—which are harder—from a place of strength. This is what M was doing in our conversation: connecting back through time with what he'd received from the living world and the people in it, and his gratitude for their care.

Now is a good moment to revisit what you're glad of and grateful for, as Macy recommends, and what gives you purpose and confidence. Knowing where those feelings come from can help you find other sources for them if climate change makes it impossible (or unjust) to find them where you find them now.

Re-Sourcing

This practice is based on knowledge and potential you already have, sources of satisfaction you already know, ways you're already good enough. It can both steady you for the next move and open you up to it.

QUESTIONS

Where does your sense of purpose and confidence come from? (Which activities, dreams, possessions, relationships, plans . . .)

How is your sense of purpose tied in with your expectations of the future?

How is it tied in with what you do right now?

How is it shaped by what you've received and/or what you want to pass on?

PRACTICE

Write, type, or voice-to-text a list of the things you do regularly that give you purpose and confidence. Go back through and underline, highlight, or otherwise mark the things on the list that feel like they're important or useful in the context of climate change. Go back once more and put a star next to the ones that are also a pleasure to do *while* you're doing them.

Groups can do it this way: Each person makes their list; people pass their lists to the left, and their neighbor underlines items for skills relevant to climate change. Then neighbors pass *back* to the list-maker, and the list-maker adds the stars of present pleasure. (If this is too labor-intensive or logistically distracting, or incompatible with any group members' abilities, just talk about it together!)

The next station in the Work That Reconnects, *honoring our pain for the world*, also requires making that pain known to each other, even when that means accompanying one another to an extreme place. The climate booth turned into that station, sometimes, for some people, like the guy I sat on the ground with. We can also find purpose in our grief itself, by making it into an active refusal of the forces that are robbing us of what we love.

In 2016, writer and geographer Maya Weeks held a memorial for the Arctic ice cap in the Ohlone territory of Huichin, which includes Oakland, California. In her invitation, she wrote, "As we anticipate losing year-round sea ice as soon as 2018, we are taking this occasion to gather and process our feelings about this changing ecosystem together. We will gather to say goodbye to the sea ice algae and the Arctic cod and the polar bear." Arctic sea ice sustains ecosystems that exist nowhere else. It also regulates the Earth's total temperature (by reflecting sunlight back into space) and holds waters that, once melted, increase sea level rise. California is, of course, a coastal state. It wasn't a big gathering, Maya told me afterward, but people came, letting the wax from the candles of their vigil soak into the pavement.

There's a tendency I sometimes hear (at the booth, on the internet, in conversation) to skip straight *to* climate grief. Most of the places, people, and relationships put at risk or under strain by climate change are still here, offering us much sustenance and joy, and needing our care and defense *more* because they're struggling. This next exercise allows you to put into practice the love for the living world that is a component of your grief while the beings you love can still benefit from it. And it's not the only way to put love into practice, or to bring the future we want—full of purpose, full of care—out of the present we have.

Celebrating Life

Maya Weeks, Diane Exavier, and Angela Blanchard contributed to this exercise. Some of the relationships, skills, considerations, and forms of organization that you'd need to do the practice in real space and time are waiting for you in upcoming chapters. It's worth imagining your way into it even before you have the skills, though, because it's much harder to do something that requires planning if we can't imagine it.

QUESTION

Where, in or near the place where you live, does ecological damage move forward quietly and with little or no opposition?

PRACTICE

Plan a big, unavoidable funeral for someone or something that climate change and/or ecological damage have taken, or are expected to take, from that place. Whatever the funeral holds—music, dancing, signs, marching, art, noise, costumes, ritual wailing—you want it to release and reverberate emotion at this loss. Ideally, you'd also place it in the way of some destructive activity. Examples might be blocking the driveway of a polluting plant so that work can't go forward, or interrupting a vote for an environmentally destructive change to laws or regulations.

Consider times of day, how people will get to the site, where they'll pee, how you'll reduce COVID risk, how physically and cognitively demanding it will be, whether and how people with a range of abilities will participate, whether and how police or security are likely to respond, and whether that's something you want to wait for. Think about roles for people who can't or don't want to be present at the funeral itself: building art, making signs, writing a press release, spreading the word on social media, watching participants' kids. Consider, too, how rage, love, and joy, as well as mourning, might form a part of the funeral.

If you're really going to do this as well as plan it, you'll need to do it with a group, to make it big and unmistakable. Pay attention to your feelings and needs, and the feelings and needs of others, as you work out the logistics. Even thinking it over will allow you to practice working through obstacles and settling on an inclusive plan.

The Work That Reconnects has *seeing with new/ancient eyes* as the third station on its Spiral: "We can sense how intimately and inextricably we are related to all that is," Macy writes on the WTR website, recalling Deb Krol's reminder that humans are another part of the Earth's way of taking care of itself. "We can taste our own power to change, and feel the texture of our living connections with past and future generations, and with our brother/sister species. . . . Then we *go forth into the actions* that call each of us, according to our situation, gifts, and limitations." Sometimes those actions are big, loud, and splashy. Sometimes they're small.

BUILDING THE FUTURE WITHIN THE PRESENT

Since he was around a year old, Conor and I have been spending one or two afternoons a week together while his mom, my friend Amber, picks up groceries or has a singing lesson or balances the family's budget or enjoys a couple episodes of TV. Before he could talk, we walked around the block at his pace, watching bees collect pollen and moss grow in sidewalk cracks. In 2022, he was six, and his sister, Maria, was four. (The three of them chose their own pseudonyms for this book.) We sang "Love Shack" on the way down the hill to the park, stopping every few feet to look at the life of the city.

When I'm with the kids, the present fills my attention. The future hovers: when I name a mullein plant and we touch its fuzzy leaves, I hope they'll remember what it's called, and that knowing the name will be a pleasure to them, a reason to respect plants, a base for further knowledge. It's similar to my reminders that we don't touch other kids on the playground without receiving permission: I'm hoping both kinds of respect and care will become part of how they behave as older children and grownups. But these things are also good *now*, or a few minutes from now, or next time: however many times they stand face-to-face with another living being. In these afternoons, the present I want is also the future I want: where sharing time, care, and interest in the world is normal, widespread, and not limited to blood family. I want people to be able to call on each other when we need a change or a rest.

In *Emergent Strategy*, her book on organizing ourselves for well-being, adrienne maree brown writes about what she calls *fractal* methods: doing small what you want to do big as a way to experience and carry out systemic change. On the day-to-day level, Amber gets a break and time to herself; I get to enjoy the kids' ever-changing company, and pass on the gifts of noticing and naming that my mom gave me; they get to (*gently*) touch bugs and parkour off their neighbors' stoops. As we build up the pattern, Amber can return to her children refreshed and with more patience. They get to learn from and be loved by a wider range of caring grownups. What I'm doing with them isn't parenting, but it helps me not miss it.

In the long run, we grow together. We love each other more, and the world more. We are more ready to live in it and to work with it. "Our friendships are systems," adrienne maree brown writes. "Our communities are systems. Let us practice upwards."

Four years after the funeral for the Arctic ice, a massive piece of Canada's last major ice shelf, the so-called Milne Ice Shelf in Nunavut, collapsed into the Arctic Ocean and its pieces began to swim away. In September 2020, the Antarctic spring, two major glaciers there—the Pine Island glacier and the Thwaites glacier—started tearing loose. This is happening here, wherever here is for you: in the weather patterns, in the closeness of the sea. The ocean loses ice; you might lose a month of necessary rain, a kind of fish you traditionally eat, a bird that used to take out a bunch of your mosquitoes on its migratory path, more money on your flood insurance, your grandparent to heat stroke. My city's creatures are tough, but how many of them will still be here in three years? In five? How can I help the kids I love feel a responsibility of care for the plants and birds and bugs we love together, without drowning them in terror and sorrow? How can I work to make for them, and later *with* them, a world that's livable in more ways than the one that we know now?

"My daughter is not going to have a choice," Dezaraye Bagalayos said against a backdrop of Eleanor's drawings and paintings. "I do tell her that. I tell her, 'You will spend your life learning how to heal the planet.'"

"What does she say?"

"She says, 'Okay. Okay.' 'Cause she talks about it, she's like, 'Our planet is sick, Mom. We made it sick.' . . . And Mike gives me these 'you're going too far' looks, but I told him, 'This can't come as a shock to her.' I am conditioning her brain to be prepared. I'm not filling her head with like, projected horror stories or anything, but I am letting her know that this will be a huge part of her life."

By loving Conor and Maria, I open myself up to one of the very things I was trying to avoid when I decided not to make a child: the deep, worldwide uncertainty of children's futures, and the pain that might be waiting for all of us there. Part of being with children in this time involves letting go of the notion (an illusion in any case) of a safe or perfect life for them. And being with them also involves preparing them—lovingly, intentionally—to weather great changes, preparation that we may not have received ourselves. What Dez wants is for Eleanor to have *always known* that life around her is under strain that's getting worse, so that even though she'll have to struggle in the reality, she won't have the extra work of accepting the idea. She's also letting Eleanor know that their family can share and face this truth together, and connect with others—of every age—to tend and share the living world more justly.

Holding Out a Hand

None of us are entering the escalation of climate change alone. In *Generation Dread*, Britt Wray writes, "Children long to be talked to in a tone that makes them feel securely attached and deeply supported. Conversations where parents are able to say, 'You're right, it's not going to be easy, but we will find a way together to manage this,' are powerful opportunities for connection." Such conversations can also be opportunities to invite younger people to join you in your caring and purposeful response to climate change, and to help build a world in which more of us can live well.

QUESTIONS

If there are children in your life, do you talk with them about climate change?

What do you say?

If you don't talk about it with them, why not?

What do you show them about how you feel?

PRACTICE

Describe—even if you're alone—what you know about climate change as you'd describe it to a child able to ask questions. Notice how this feels, even if it's not your first time saying the words. Say them a few times more, making adjustments until you arrive at a version that you're willing to try (or try again) with actual children in your life. If you're having an emotional reaction that feels like you'd need to keep it out of a conversation with a child—too explosive, too bleak—you may need to revise what you're saying. Or it may mean that you *must* say it but need to pause first and use a steadying or grounding exercise—something you could invite a child or children to join you in.

If you're doing this as part of a group, do it as a partner exercise: one person speaks, and the other person then speaks *back* what they said, keeping the words as close to the same as possible. Help each other make the version of the conversation that you're willing to try beyond the group, and support and encourage one another before and after—maybe making a phone or text date to check in.

If you or your child don't (or sometimes don't) process spoken language, you have your ways of sharing information and helping them handle stress and fear. And there might be wrong times for the actual conversation, like a moment when another distress or transition is filling their attention or yours. But there isn't a "right" time to talk about climate realities together. There isn't a "good" time. There is only the time we have now.

BECOMING A GOOD ANCESTOR

Early September 2019 and the afternoon felt too warm and damp, feverish. The kids and I were in their family's backyard. Maria, around two then, was a little droopy, and I held her on my lap on the swing, just rocking, while Conor dug industriously in the dirt under us. If it started to rain, we could go inside. If one of them had to poop, the plumbing was working. There were snacks in the fridge, and the fridge was running. Whole, quiet, everything we needed near to hand: of course I wanted to keep us in that moment. And of course I couldn't. Trying to do so—to push away the realities of change, to freeze my sense of a good present or even a good future for them—would ultimately hurt them and the world we love together, the people we know, and the people we will never know. Climate breakdown means that the shape of even our good days together is going to have to change, and we must look for the places where we can direct that change.

Two years after we met, Dezaraye Bagalayos began working with the Allensworth Progressive Association to bring the town's original vision, a self-sustaining Black community with a focus on education and industry, into the present and future. She says it's "the work of a lifetime" and "a balm to my soul and my climate grief." Dez's students, many of them the children of farmworkers compelled by their bosses to deplete the land, are now learning how to practice regenerative agriculture—which restores the land's health and works with its tendencies—and how to organize and operate a workers' cooperative, so that their labor and its results will be their own, and their decisions will be shared. The history of land and anti-Blackness, colonization and displacement, enslavement and sharecropping, violence and broken promises can't be fully healed by one small town. In the 2023 floods, the wound was renewed when industrial agriculture companies and the railroad company diverted floodwaters toward Allensworth rather than their own lands. But the town and the farm are sites for transformation of history, and of possibility.

"Our overall goal for Allensworth is for it to... actively demonstrate how Central Valley rural communities will not only have a chance to survive climate change, but also thrive and provide alternative examples of how we can better live on the land," Dez wrote to me. "Being able to share what I do with Eleanor is one of the best parts of my job, showing her the kind of work more and more of us will be engaging in as climate change continues to make its realities known to us. I need my daughter to know that even though we may be up against seemingly impossible obstacles, we have it within our soul and spirit to meet those challenges, and we have an obligation to ourselves, those we love, and the planet that gives us all life to engage in that work and cultivate joy in the process."

Driving between the one-crop fields through the Tulare Lake basin that's been his life's longest knowledge and work, Dezaraye's teacher Rob Hansen pointed out valley oaks lining drained watercourses. "The dead trees are totally us," he said, *us* meaning humans and, more specifically, California settlers' agriculture industry. The biggest of the channels on the nearby Kaweah Oaks Preserve had no water flowing through it for five years in a row. "Some of the oaks came through one year, some made it through two with some stress, but in years three to five, boom"—100-, 150-year-old trees were dead and standing wood, no longer anchoring carbon dioxide or providing oxygen, shade, or habitat, and fast losing their ability to anchor soil. The upland ecosystem in a freshwater sea of marsh, where they'd evolved to grow, was gone, and the groundwater that would have helped them endure had been sucked away and hoarded. But even before the wetter winter of 2018–2019, there were a few trees whose consistently returning leaves seemed to be saying, *What drought?*

"Out of all the millions of acorns that those trees have dropped over the years," Rob said, maybe those few had "the right stuff" to survive under conditions like these built into their DNA, so that they didn't become part of the surrounding ghost forest. "Maybe they get more of their water from surface roots instead of a taproot seeking the now-absent or deeper groundwater. We may need to give up on that kind of streamside oak forest in the Valley, we've changed it so much. We may just be going through a bottleneck, where the trees that evolved dependent on that pre-climate change hydrology will disappear, to be replaced by some kind of modified, more drought-resilient forest."

As Don Hankins reminded us in Chapter 3, it is possible—though not guaranteed—for many generations of a people to ride out great changes and losses. How we do that, who we are while we're doing it, how we're teaching each other to be: those are things we can control within uncertainty, sorrow, even horror. Our dreams and plans, our desires for certainty, served us for a while; we don't need to feel bad that we had them in order to lay them to rest and open ourselves to new ones.

Planting Your Thanks

Here's an exercise to practice farewell and prepare for welcome. We can mourn that future you craved, that version of you, those things you value. You can value them and still let them go.

QUESTIONS

What's a dream of the future you're ready to let go of, now?

What did it offer you while you held it?

What will you be free to do when you let it go?

PRACTICE

You'll need these things:

- A tree or shrub seedling, ideally one that evolved where you live
- A rotten vegetable, fruit peel, etc.
- An area of ground or soil where a tree or shrub could grow

- A shovel
- Water and something in which to carry it
- Someone to help dig if you can't dig
- A dream of the future you are ready to let go of

Whoever's digging, dig a hole in the ground you've chosen, working around any existing roots. Thank the ground for bringing us forth and receiving us. Hold the fruit or vegetable corpse, and thank your old dream of the future for doing what it did for you—stimulating you, sustaining you. Place your dream of the future in the hole. Pour the water over it, and as you do, forgive your dream of the future for doing whatever you wish it hadn't done—distracting you, exhausting you, or just not being possible. Forgive yourself.

Place the seedling in the hole and spread out its roots. Fill in the dirt and pat it down; pour in a little more water. Ask your old dream to nourish the seedling; ask the seedling to transform the old dream into something more livable, flexible, and strong. You'll need to tend the tree as you would any seedling, but if you have to leave it, know that you gave it the best start you could.

I'm not asking you to open your hand for emptiness, but to free it for effort and connection. When we give up one thing on purpose, we make room for another.

CHAPTER 5

ON COMPANY TIME

FROM THE BOSS'S PRIORITIES TO YOURS

YOU ARE HERE

This chapter invites you to do these things:

- Explore the connections among your work, your sense of self, your climate anxiety, the forces that cause it, and the rest of the living world.
- Learn more about *Just Transition* and *degrowth*: what they are and how they might shape our climate present and future.
- Reassess what satisfies you and what feels like "enough" to you, even when you are also dealing with or fearing scarcity.
- Look for the *openings* and *levers* for change in, or around, or beyond the way you earn your living.

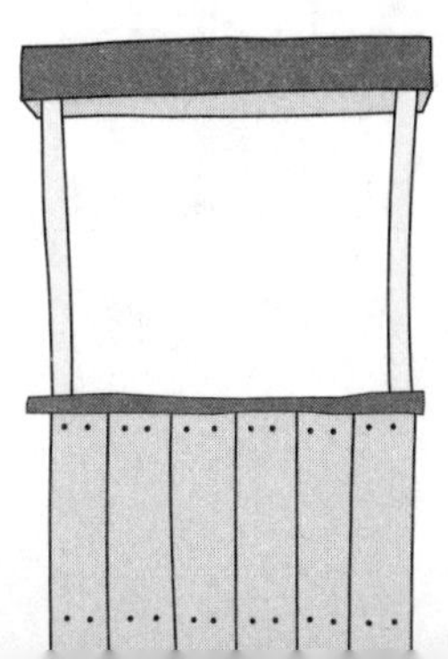

WEAKENING THE TRAP

You've probably seen people writing and talking about *divestment* from fossil fuels. Students demand that universities reject funding from mining companies; organizers will talk you through switching banks if your bank lends money for building oil pipelines. The goal of divestment is to weaken the institutions and systems that force climate change—to make it less profitable for them to do what they do, and also to give them less control over what *you* do.

Visiting my best friend in North Carolina, I got talking with her boyfriend, Nate, who works for an asphalt and paving company. He relayed a story of newly paved roads washed away in flash floods—because the company was cutting corners, the preparation of the surface would have been inadequate even in ordinary conditions and was more vulnerable to increasingly extreme weather. The connection seemed clear, so I asked: "Does anyone ever bring it up, that climate change is why this is happening? That it's affecting your work? Like, what would happen if you brought it up?"

"They'd be like, 'What the fuck you talking to me about that for.' What you have to understand is, we have to do work that doesn't make sense all the time." He gave another example: a driveway that another crew laid down, only to be immediately told to tear it up again. They knew that it wouldn't work in that terrain, but had to follow instructions—both times. They have little control over which jobs they do, where, and how. The way they earn their living and practice their skill is vulnerable to climate change; it's also a hot, poisonous, and tiring job that uses a lot of petrochemical materials. And the knowledge that they've developed about what will and won't work is often ignored by the people who do have control. Nate recognizes climate change as a reality. He knows the effect it's having on his work. But there is no clear path for that knowledge to purposefully change the way the work is done.

To be invested in a thing—a system, a relationship, a place—means that something of yours is tied up in it. Sometimes that tie is a bond. Sometimes it's a trap. In the first three chapters, you practiced strengthening ties that are bonds. In Chapter 4, you began a process that continues in this chapter: weakening the ties that are traps, even if we can't afford to break them yet.

Capitalism, whose priorities are changing our climate, has us all invested in it to some degree. But I don't buy it that we can only survive by staying exactly as invested in it as we are today, while it continues to squeeze us dry. I don't buy it that there's *nothing* different you can do, no move you can make. Which aspects of the system or yourself you move—where you can push, what you can weaken, and what the costs and benefits will be—are particular to you, although when you look at them with others, you may find that they are also shared. This next exercise will help you identify some of the *openings* (opportunities for change) and *levers* (places where pressure might create change) for responding to climate realities in a part of your life that often feels especially immovable: the way you earn your living.

Letting Yourself Go

This exercise works differently than the others: you'll be able to see more about the relationships among your work and the various things you get from it if you do the practice first, then ask the questions.

PRACTICE

Write a budget—with two columns like a regular budget, expenses and income—for what you give to your work (paid and unpaid) and what you get from it *in the present.* On the "expenses" side you might put things like time, attention, physical effort, mental effort, patience. On the "income" side, besides money (if any), you might put things like pride, skills you've developed, friendships, endurance, perspective. When you've made your lists, highlight or underline the things that are most important to you—what you're eager to keep or reach, or afraid to lose. Save this budget. You'll use it in another exercise later.

If you're doing this as part of a group, talk and share as you go—learning what other people do and don't get out of, or give to, their areas of work will help you know each other better. Later on, you may also have suggestions for one another about how to get the things on the "income" side in other ways.

QUESTIONS

How does the work you do or have done for money (if any) also shape your sense of self?

If you don't or can't work for money, how does *that* shape your sense of self? (This doesn't mean you're not working!)

How do the answers to those questions (the ones that apply to you) shape your ability to respond to climate change?

Some people find this one frustrating, for several reasons. Maybe all that paid work offers you is financial survival—if that. Maybe your work, paid or unpaid, isn't the direct source of your sense of self or purpose at all, but just a way to *get to* things that matter more to you, or to be the kind of family member, artist, athlete, religious participant, etc. that you want to be. Or maybe the last question is the frustrating one, calling attention to the aspects of your work that interact with climate change in ways you don't control.

DOES IT HAVE TO BE LIKE THIS?

Before Angela Blanchard was the CEO of Houston-based community development organization BakerRipley and a disaster expert by profession—she works with people and governments after catastrophes, from wildfire to war, to meet needs and develop collective strengths—she was a disaster expert by experience. She's from the Gulf Coast on the Texas side, and she reels off the names of hurricanes she's lived through like the names of cousins. Allison, Katrina, Rita, Ike, Harvey, Ida: that's just the youngest generation.

The area of the United States where Angela does much of her community development and disaster response includes Cancer Alley, the land along the Mississippi River between New Orleans and Baton Rouge. Over thirty petrochemical plants operate there, and companies are building more. Longtime residents, mostly Black and working poor, report one to two cancer deaths in every ten or so households—nearly fifty times the national average in at least one town—as well as an increase in miscarriages and respiratory illness. An elder interviewed for *Pro Publica* in 2019 recalled that when the industry moved in, "we thought we'd get better jobs, but [the plants] brought their own people here."

The source of Angela's homeland's suffering is also the source of its wealth: Texas crude, the processes for extracting, refining, and transporting it, and the labor of the people who carry out those processes. The people directing those plants rely on them not just for profit, but for their senses of self. They won't stop what they're doing without major legal or regulatory pressure—or without a massive drop in supply, demand, or labor.

Angela doesn't see any of that happening. "If we want to do away with fossil fuels," she wrote in an early draft of the book she's working on, "we give up controlling our indoor climate, we stop eating anything not grown locally. We stop selling anything in a plastic package. We stay home—wherever that is. We get used to being hot, dirty, thirsty and hungry. That's life without petrochem and fossil fuels. Hot, dirty, thirsty and hungry. And that is also life with fossil fuels."

Angela meant both the lives that so many people, especially impoverished people, are living now, and the lives that all people can expect to live in the future. She's said to me in conversation, more than once, that the only thing worse for humans' quality of life than continuing fossil fuel production would be stopping it completely. This is where a lot of public *and* internal—internalized—stories of climate change leave us: trapped in systems that are hurting us, easing our breathing through a tube whose manufacture makes our breathing more difficult.

When people point out how vile the present is, as Angela did, and express their confidence that the future will be the same or worse, it's possible to receive that as a reason for despair and inaction. We can also train ourselves to receive it as an aid to letting go, in order to make room for something better.

Steadying the Weight

Relieving the pressure of the trap momentarily, through the signals our bodies are sending, helps us gain more emotional space and balance to figure out how to pry the trap open thoroughly and permanently. This is a grounding exercise that you can do anytime, and like the other grounding exercises, it works better if you can make a habit of it.

QUESTIONS

Where are you, typically, when you feel the pressure of the trap: how these economic systems limit you, how they grind you down? (At work, in bed, at school, in a medical waiting room, up the side of a mountain... You can choose more than one place.)

What do you usually do when (or right after) you feel it?

PRACTICE

In one of the places where you feel the pressure of the trap, deliberately allow yourself to feel it. Then inhale and stretch as big a stretch as your body will permit. Don't hurt yourself, but push right up to the point where it's about to be uncomfortable. Exhale, relax, and on the next inhale see if you can resume your *knowledge* of the trap without the trapped *feeling*.

If you're doing this as part of a group, choose partners or trios and coordinate your stretches—in the same place if possible, synchronizing your watches if not.

Physically moving and watching other people move can loosen up our minds when we're feeling stuck in our situations. It doesn't change the situations themselves, but it allows the question of whether and how we want to act, and what might be possible if we do, to become a real question again.

FINDING OTHER WAYS TO GET WHAT WE NEED

Instantly stopping all fossil fuel mining, transport, and processing is not going to happen, so it's silly to make scenarios about that. It's not silly to think about needing those industries less, and about the other changes, at all scales, that might make that possible. But it is difficult to think about, and difficult to feel, when these are statements that a lot of us can make truthfully: "But I have to work to live!" "But my job makes the climate crisis worse!" "But my industry is impossible to change!"

But your industry is changing—is being changed. The pavement is sliding away from the ground. Bridges and train tracks are buckling in the heat: if you sell a product that's made elsewhere, or commute to your job, that's going to matter. In 2012, midwestern US corn crops were decimated by drought; in 2019, they were rotted out by floods. If you work in a restaurant or raise livestock, that's a hit you feel. Longer and hotter wildfire seasons have changed every kind of outdoor work where they take hold, from farm labor to hiking and kayaking tours; flooding changes not just construction but insurance. Almost all of us who currently work do so within systems that were developed before the world started warming. Unless they are transformed, what they now provide in exchange for our effort and skill (and in many cases, our health, freedom of movement, and relationships) they may not be able to provide any longer.

As you know, those systems have plenty of problems beyond their fragility. It would be a great relief to leave them behind, as long as there are better replacements lined up. And there are starting to be.

The Just Transition framework, which I learned about through the Climate Justice Alliance (CJA), shows one way that a combination of policy, investment/divestment, and labor organizing might bring about an economy that's more livable for more people. It outlines a transition from the fossil-fuel-dependent economy that Angela Blanchard described, from other extractive industries (mining, clear-cutting, some kinds of farming) and from those and other industries' dependence on low-waged, unpaid, or slave labor.

The transition is toward ways of meeting needs—as consumers and as workers, as humans and more-than-humans—that are either neutral or healing for living systems and the people in them. The CJA states on its website, "Just Transition initiatives are shifting from dirty energy to energy democracy, from funding highways to expanding public transit, from incinerators and landfills to zero waste, from industrial food systems to food sovereignty, from gentrification to community land rights . . . and from rampant destructive development to ecosystem restoration. Core to a Just Transition is deep democracy in which workers and communities have control over the decisions that affect their daily lives."

This emphasis on worker and community control comes from the labor unions and environmental justice groups who collaborated to describe and build power for this shift, in part as a way to keep workers from being stranded if and when environmental policy affects their jobs—what will plant workers do if the petrochemical plants in Cancer Alley close because they can't meet new regulations? What will Nate and his coworkers do if people stop paving their driveways, or fewer companies maintain their parking lots? Because of its origins, the Just Transition framework also addresses the question of how you'll eat and stay sheltered on the way from here to there, by calling for the creation of new, useful paid work that is (or could be) waiting for people when they leave their old, destructive work. Just Transition principles backed by a combination of public and private funds have set up and shaped tribally run and staffed renewable energy on Diné land, and subsidized opportunities for industrial farmworkers in Florida to use more sustainable farming techniques. The regenerative, cooperatively owned farm that Dezaraye Bagalayos, her colleagues, and her students are building is another example. These endeavors don't need to be huge to show us how they work.

Know Waste Lands, the site of BK ROT's compost hauling and processing service, is the size of an ordinary lot in Brooklyn. High brown piles of compost in the making are visible through the fence, a mix of food scraps and wood chips turned regularly by employees and volunteers to advance the transformation into healthy soil. Some of the food scraps are dropped off by people strolling in with full bags and little dogs on-leash. Some get picked up by Victor Ibarra, who hauls them from restaurants and cafés in his bike trailer. In 2019, he'd been working with BK ROT for six years. A lot about the organization and the neighborhood had changed, he said, but his main job was still collecting food scraps, processing them, turning them into compost. He lived in the neighborhood, and his schedule had room for him to pick his little brother up from school.

This is how BK ROT plays their small role in a "multilayered solution" for garbage in New York City, which when I interviewed Victor was producing 145,000 pounds of food scraps every day. BK ROT is a segment of a Just Transition because it's set up to sustain not just soil, but people. The haulers and processors are mainly young people of color from the neighborhood, paid fair wages for their work. Founder Sandy Nurse and her colleagues organized it from the ground up, so to speak, to feed into justice at every level. "We're going for systems change," Victor's colleague Renée Peperone said. "It doesn't matter if people agree with our politics. People are contributing regardless because of the way it's set up. [With composting], if you do a little or a lot, you're still contributing and you can see it right away. If you're just dropping off your compost every week, if you're paying for compost, it goes toward paying youth of color. That's who we center in all our work. You can't do the environmental part without all the rest of it."

Through BK ROT's pay structure and hiring priorities, some of the people hit hardest by capitalism and white supremacy feed themselves, their families, and the soil. No systems are perfectly climate-change-resilient, but their main activities (biking, digging) and their main materials (food scraps, wood chips) are fairly adaptable to changing conditions. And while the work *is* work, it's laying the groundwork for a little more abundance in the future, rather than making that future even harder.

Nearly all humans spend at least some time being hurt by activities that also allow us to survive: our paid and unpaid labor. The previous chapters reminded us that our lives on Earth are rich in the beauty, variety, pleasure, and grief of being alive with others, and that we have the power to love them by contributing to their well-being and ours at the same time. What would it be like if we could do that through our work?

Soon after the 2022 IPCC (Intergovernmental Panel on Climate Change) report came out, Rebecca Leber, who covers climate change for *Vox*, wrote about the report's recommendations for changing the systems we use to live, eat, and move around, in order to lower greenhouse gas emissions. She identified some areas of professional life where that's especially possible: "Architects' choices can mean the difference between new buildings reliant on heat pumps [powered with renewable energy] or gas-powered appliances. . . . Landlords can help ensure that rentals are energy-efficient, keep appliances well maintained, and adopt clean energy use, all while decreasing energy costs for the building. City planners can make roads safer to bike and walk, and discourage road traffic, all through smart design."

They could, but why would they? Out of the possible reasons, two stand out that might be relevant to you personally: they might receive financial incentives through policy that you could push for (funds for making changes, penalties for failing to make them) or they might respond to organized pressure (from employees, colleagues, tenants, and clients) that you could contribute to. In the second example, you could link that pressure with demands for higher wages, better working conditions, or other shared needs. There's also potential for more mutual and informed relationships between people who (for instance) might be making decisions about where the pavement goes, and people like the members of Nate's crew, who have expertise that could inform where it should go. Here are some questions to ask about your own job, field, or industry, to see if any levers or openings for greater climate responsiveness are present there.

Developing Your Career

QUESTIONS

How does the work you do, paid and unpaid, interact with the land, water, and air where you live? Look in both directions: how it affects living and built ecosystems, and how it's affected by them.

How about its interactions with the land, water, and air elsewhere?

If it's paid work, when there are changes in your industry or field, why do they usually happen? (Government policy, new technology, labor organizing...)

If it's unpaid work, when there are changes, why do they usually happen? (Rise or fall in available income, time opens up, someone's abilities change...)

PRACTICE

Return to the "income" column of the budget you wrote in the practice for "Letting Yourself Go" beginning on page 101.

Use it to write or tell (or draw!) the story of *one* way to do your job differently that's more compatible with the health of the world than it is now. It can include skills and knowledge you developed because of your working life (or other aspects of your life); it doesn't need to be a version of your current labor, paid or unpaid; for the purposes of this practice, you're not required to imagine where the money to make the change would come from.

Now ask *who else* that change could benefit, and how: You and the people close to you? Coworkers or people with less power than you in your workplace? People who are served by the job? Imagine *one* more adjustment that would allow the benefits of this change to go even further, creating a reason why another group of people might favor or even work for the change with you.

Existing groups beyond the workplace can do this one together. But it will have more potential as a point of entry for action if you also do it with one or more coworkers, colleagues, or people you live with.

Keeping a transition just is an additional challenge. A Diné student of mine agrees that the solar project is a good idea, but she's witnessed that it's slow to grow; in her opinion, there's a combination of reasons having to do with cost and trust. Know Waste Lands is in the Bushwick neighborhood, and Bushwick is being gentrified; Renée acknowledged that most of the people who volunteer and drop off their food scraps are white newcomers. That disappoints her, although she finds it appropriate for them to be involved in repairing some of the environmental damage that white supremacy has caused, and letting their higher income flow toward the sustenance of their Black and brown neighbors. All our efforts toward Just Transition are going forward in a world that's shaped by our environmental history, which includes colonization, enslavement, and exploitation, and is driven by inequality. To invest in Just Transition, we need divestment from all those things, at all levels: business and policy, and also—where they're a factor—our senses of self.

BUILDING AND UNBUILDING AT THE SAME TIME

Food turns into waste turns into soil turns into food. Economist Herman Daly pointed out in a 2022 interview with *The New York Times* that while the Earth's living systems are more like BK ROT, the dominant economic systems at present (and for the last few centuries) move only in one direction: digging more stuff out, using more stuff, ruining more stuff. Continuous financial growth depends on activities that kill living beings and places by turning them into stuff, and it's the major driver of climate change.

Humans should align our economic systems with our ecological realities, Daly says. So do other people who study, propose, and in some cases try to enact *degrowth*, a planned and democratic reduction of production and consumption. Degrowth is making less, using less, and doing less—not just at your job in order to get some time to breathe and think, but as part of a transformed social and cultural pattern.

Degrowth, for a region or a country that applied it, would mean that businesses stopped trying to make more money every year—continuous financial growth would no longer be expected (or, in the case of corporations, demanded). It doesn't explicitly refer to workers owning what they make and do, rather than selling it to their bosses, but that's probably the likeliest way to achieve it. It would mean manufacturing fewer objects, and being more selective about which ones: maybe oxygen lines are a go but yogurt containers and plushies are over; maybe wheelchair vans are common but F-450 pickup trucks are rare. It would mean generating less electricity, traveling more rarely and slowly, changing our food systems as well as our diets. In fact, it's got some things in common with the low-fossil-fuel world Angela Blanchard was describing earlier in the chapter. A little hotter—but it's going to be like that anyway. Limited travel—but the roads are already melting. Tired people—but how many of us can honestly say we're not tired now? If our lives really are going to be harder no matter what, why not put that difficulty in the service of a more lively and various world, where work and well-being are more fairly distributed?

Throughout this book, I've tried to share stories of more livable systems and lives, so that you know what can be done and can imagine even more. But there aren't that many stories of degrowth yet, because it scares us. Growth as an essential of economic life is so ingrained in the structures that affect us that the idea of doing, earning, saving, gaining, and keeping *less* may seem ridiculous, threatening, or irresponsible. Think back to your dreams, visions, and stories about the future in the previous chapter: How many of them depend on your personal economic growth, and how entangled are they with the economic growth of companies or countries? Like a Just Transition, degrowth needs to happen on many levels—policy, business, personal—if it is to be effective.

In a continuous growth economy, those who grow more, use more, buy more, and burn more do so at the expense of poorer people and colonized people; they outsource illness, injury, and disability; they lay the foundations for future disaster after future disaster. Our current economic systems and those who benefit most from them tell us that divesting from them is dangerous: that what we have could be ripped away from us (by them) at any time, that we always need more (from them) in order to protect ourselves (from them) or to recover from the suffering inflicted on us (by them). This *feels* equally true across the board—but it isn't. The better you're doing financially, the more our current systems serve you, the likelier you are to be invested in them. And the more able, and more obligated, to divest from them you're likely to be.

Alongside efforts for the greater material equality offered by workers' rights, worker ownership, and a Just Transition, there is inner unbuilding about money, safety, and pleasure that we can do to help us prepare and commit to the outward reality of degrowth, which international climate scientists say is key to the survival of the living world. Caroline Contillo, the climate trauma social worker who offered us our very first practice in Chapter 1, notes that the inner antidote to endless growth (and depletion) is recognizing satisfaction—how it feels in your body and mind. She recommends finding and welcoming the *feeling* of sufficiency, of enough-for-now. This doesn't mean being content forever with *not* enough, or accepting impoverishment in service of some vague climate goal or utopian gimmick. It means gently but genuinely asking yourself about the overlap between what you need, what you have, and what you strive for—and where your own openings and levers for change might be. And you can only answer those questions if you're able to recognize and experience the times when your needs are being met.

Touring Your Blessings

Because the economic structures that we're all invested in to some degree depend on the fear as well as the fact of not having enough, recognizing our satisfaction when we find it can help us to give those structures the bare minimum, and weaken the hold they have on us.

QUESTIONS

What's one thing you like, enjoy, or are grateful for that you're also rich in? (Houseplants, bobbleheads, songs memorized, opportunities to cook food you like, places you've traveled . . .)

If you want more of that, what do you think more would offer you?

PRACTICE

Set aside a small timespan—say ten to fifteen minutes—to spend luxuriating in your answer to the first question. If they're objects, touch each one, appreciate its qualities. If they're living beings, pet or tend them. If it's a process like swimming or cooking, immerse yourself in it for the stretch of time you choose. If it's places you've visited, look at your photos. At the end of the timespan, formally thank this thing or process for being in your life. You could try saying, "You're enough for me," even if it doesn't feel true, just to hear how it sounds.

If you're doing this as part of a group, you can give each other tours (or concerts, or whatever format makes sense) of the things you like, or have a "show and tell" where each person shows off a sample. You can also call on each other to help create the fifteen-minute timespans by taking on care work or other chores.

That feeling of your needs being met and your pleasures being present? That's the feeling I want everyone to have, accurately, more of the time! We know that both what we need and what we enjoy are threatened by climate change and the forces and systems that cause it. The people who read this book are not *coming from* equal degrees of investment in those systems, and so the ways you divest will be different. Those who've been more heavily burdened by the forces that cause climate change should not also be more heavily burdened by the work of fighting it, though that's often the case today. But where we *move toward*—through some giving more and some giving less, some giving up more and some giving up less—that is a place where greater equality is possible.

MAKING TIME FOR THE WORLD YOU WANT

Victor Ibarra said honestly that hauling and processing for BK ROT is hard physical work, and the smell of rotting may be too much for some of the young people they're looking to employ, teach, and support. But he believes the work gives back as much as it takes. "It's something that people should be learning," he said. If a small number of people can do this much, what happens when a whole city starts doing it? "Every single lot that is empty right now could be something like BK ROT." In addition to the pay, BK ROT offers people who work for them this sense of purpose and possibility: they know they're working to improve a situation that's currently harmful to the world and the people in it.

Many of the people you've met and will meet in these pages are working people, and while some of them now have jobs that align their efforts with the efforts of the living world to live, they didn't always. Almost all of them have survived work that ground them down. Many are parents and caregivers. Many are disabled or chronically ill—sometimes because of the work they did or do, the record of environmental injustice on their bodies, or the stress of living in a racist and extractive system.

I'm absolutely *not* saying, "If they can do it, so can you." I don't know what you can do! Plenty of people in those same circumstances—which may also be yours—aren't able to, or don't choose to, shift their allegiance to work more for each other and for life on Earth, and less for their bosses. Their efforts are going toward a safer and more stable present and future for themselves and the people closest to them. They have agency, just as you do. I'm also not encouraging you to accept unfair pay or lousy conditions in the name of "doing good" for the climate or anything else. I am asking if *you* know what you can do—if you're willing to revisit these questions: Who and what do you want to work for? What kinds of work do you want to make more common, more possible?

If you're tempted to say that what you *want* doesn't change anything, that's exactly my point. Your boss, their boss, the company, even capitalism itself doesn't control what you want. Depending on your job, they may control what you do, what you say, what you make, when you eat, when you pee, when you smoke, but they cannot control what you want. For that reason alone, I would say, it is worth wanting more. Here's an exercise to open up a little space, so that you can better imagine new or different ways of working, living, and finding purpose.

Leaning, Not Cleaning

Even imagining this practice may open up a sense of possibility—and a sense of dread. Sometimes that's dread about the near future: it's a job you can't afford to lose. Sometimes it's about the further future: you're hoping that working up to or beyond your capacity now will offer more security or more opportunities for pleasure and rest. But climate change calls our whole future into question. In some jobs, doing less *is* dangerous for someone, and this exercise won't work for you. But if you can keep a job while doing less at it, the time and energy you free up become yours, and you can use them to reflect on what you might want for yourself *and* for the living world.

QUESTIONS

Where in your jobs—paid and unpaid—could you do less without putting anyone in danger?

If you do less at work, where would you like to place the energy and attention you save?

Where could you place it in ways that serve you *and* the rest of the living world?

Where do the answers to those questions overlap?

PRACTICE

You'll need a bunch of sticky notes. On them, write instructions for doing less or slowing down, for all the parts of your job where that is possible, as though you were going to put them up throughout your worksite. If they also make the job easier on your body, even better. You might write "Walk very slowly and sweep without effort or stooping" on a sticky note for the broom. A note that reads "Chop the onions a little bigger" could go on the knife. A note that reads "5 minutes MAX on any email" could go next to your keyboard. You can also write notes for your work that isn't paid, like housework in your home: the knife note and the broom note could apply there, too, and maybe the sheets can go another couple of days without washing.

If you're doing this as part of a group, you can describe your jobs to each other and suggest placements for the notes and adjustments to the tasks—it's also useful for people to know what's involved in each other's work. You can show each other love and enthusiasm by texting each other photos of the notes in place. (Obviously, don't leave them in place unless you're ready to quit—but you could try doing what they say anyway, after you remove them.)

Rebecca Leber's suggestions about certain fields and kinds of work offer a way to make the workplace itself the arena of change, most likely through organizing: uniting with colleagues in a demand to do your work in a way that's better for you *and* the planet. The opening comes from the kind of work it is, and the lever is the collective demand. You might find that your path to participating in environmental justice is through building or joining a union, or pushing the union you're already in to add environmental justice to the list of things it works toward, or learning a skill for different work as part of a Just Transition, or phoning it in at a medium-pay, medium-status job in order to share your time and energy with the living world in other ways.

Any activity that weakens the trap and strengthens the bonds—donating to a strike fund or bringing donuts to a picket line, advocating for state or federal investment in more humane and less lethal work—is activity worth doing. And it's worth asking what would make it more of a pleasure, more of a satisfaction, to live with yourself—not just in the ethical sense, but in the sense of identity and purpose, the self you're making through what you do.

Listening at the Climate Anxiety Counseling booth to people in many different economic circumstances, trapped in all kinds of work, prepared me to understand the need for a Just Transition. It directed my choices of climate and community justice efforts to support, and how to behave while supporting them. And it trained me to find fulfillment in this listening and support that I might otherwise have looked for in parenting or my professional career or my activities as a poet. At the booth, I also had the luxury of talking with one person at a time, tailoring my questions and recommendations about the relationship they had and wanted to have—to climate change *and* climate action—to what they shared with me about their particular lives.

Capitalism thrives on the difficulties it creates. One of those difficulties is that it can be tough to communicate with people who are positioned differently from you, people who've been injured in different ways or to different degrees by the forces that cause climate change and so much other damage. And yet we still can learn to be on one another's side, to redistribute efforts and sacrifices more fairly, and to recognize the potential in our shared survival. Because that's challenging and worth doing—like processing compost—we may even learn to take our satisfaction from it.

Jamie Tyberg, an organizer and scholar, writes in a paper called "Unlearning: From Degrowth to Decolonization" that "degrowth tells us to care for the Earth's systems to care for the people, and to redistribute any surpluses back to the land and the people." If you're doing any of these things, you're bringing the world we want out of the world we have, making our survival more likely and more possible the more of them you do. We all belong in the future, in some form—but maybe not in the exact forms we bear now. When the openings don't present themselves, we have to find them or make them, lean on the levers, walk through the doors, together.

CHAPTER 6

CHANGING THE STORY

FROM INDIVIDUAL MOTIVATION TO COLLECTIVE ACTION

YOU ARE HERE

This chapter invites you to do these things:

- Explore a few of the many ways to organize with others in response to climate change, its causes, and the other threats to life and well-being with which it interacts.
- Find the sweet spot where what's needed in climate and community organizing meets what you're able to do, and what feels right to you.
- Observe and reflect on organizations—and yourself—to make sure you're not reenacting harmful structures and hierarchies, but changing them for the better.

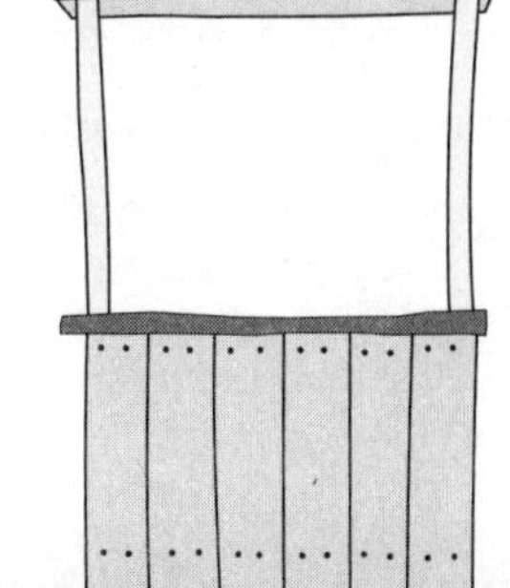

THE STORY SO FAR

Let's say that your internal and relational efforts are continuing to transform you: you recognize that you have some power to respond to climate change, power that is real even within the trap-like aspects of your circumstances, and you've freed up a little time to consider how you want to use it. If your paid work is where you want to make change with others—changes to your workplace or your industry that create more fairness and less suffering, that walk back or limit environmental harms, or that benefit the living world—then what's next for you is difficult, but clearer than it was, and may happen in a place where you already go most days. But if that's not the case, what will you do now?

The complexity that can make systems intimidating, their many parts and relationships, is also a gift because it offers many paths to transforming them. Figuring out where you fit in involves exploring what's out there to fit into, matching it with your existing skills and inclinations, and accepting opportunities to transform yourself further.

You've been involved in that process throughout this book. The identification of internal and community strengths in Chapters 1 and 2, the interspecies and intergenerational relationship-building in Chapters 3 and 4, and the reevaluation of your working life and loyalties in Chapter 5 all offer paths to participatory action: responding to climate disasters, nourishing the living world, finding the levers and openings to turn your energy toward reparative work. You've also been practicing the emotional and somatic (bodily) methods to help you do those things in ways that are more sustainable for you and the people around you.

The rest of this chapter, and parts of the next, focus on finding and building more formal group relationships for collective climate action—especially for policy, advocacy, agitation, and direct action—and on some of the situations and feelings you might encounter. The questions in this next exercise are mainly designed for people and groups who are shifting some of their time, focus, and care in that direction. In the practice, there's an option for people and groups who are already working together toward environmental justice and ecological repair.

Zeroing In

It's good for you and for the effectiveness of any group you're part of to periodically reassess what you can realistically and even pleasurably offer. Even if the cost of changing your contribution at that moment is higher than you want to pay, it is still good to check in with yourself (and each other) and push for an adjustment when you can. Your sense of responsibility to your community, your ecosystem, and yourself can lead you to ignore your limits, but the small responsibility that you carry out is better than the big promise you break.

QUESTIONS

What would be *easy* for you to supply to a group or organization working for ecological well-being? (Meeting space, rides or transportation, free access to a service you usually sell, research, phone banking, storage space, cooking, writing, testifying at public hearings, childcare, jail or court support...)

What would be *satisfying* to supply?

Would you rather make your contributions to climate and environmental justice in person, or remotely? Through gathering with others, or on your own time?

About how much time per week can you—or do you—offer if other things in your life stay more or less the same?

PRACTICE

Clear the amount of time you think you can offer most weeks, whether it's minutes or hours, calling on the questions and practices for "Getting Ready and Getting Rested" (see page 16) and "Leaning, Not Cleaning" (see page 119). For at least two weeks in a row, though, don't plan or commit to *anything* new for that time. Outside of emergencies, use it for rest and rest only (whatever is restful to you). Note how you feel at the beginning and end of each rest period, and also at the end of the two weeks: Has your sense of what and how much you can share changed at all, or is it about the same?

Groups that are actively working for climate or environmental justice, survival, and well-being can build this in as a two-week break (or drastic scaling back of work—again, outside of emergencies), after which you can take stock of participants' time and skills. Does anyone want or need to change how or how much they're participating? Is that possible now, and if not, when can you make it happen?

WHAT WILL I DO NOW?

In the summer of 2017, so many people I knew were asking themselves that question, and not just about climate change: they were looking for ways to respond to the Trump presidency and the dynamics it was intensifying in US life. Organizations that had been fighting for better living conditions and freer lives, under many other administrations, were trying to direct this new wave of eagerness and energy and resources, and also trying not to say—or to say with strategic gentleness—"What took you so long?"

In many cases, they were also frustrated by a lack of humility and willingness to be guided by their expertise. For years they'd been facing and studying the policies and attitudes that these mainly white newcomers were only now lining up to fight. These included environmental injustice and many other systemic and institutional injustices that interact with climate issues: struggles for fair and safe housing, attempts to limit police violence, freedom of movement for immigrants and refugees, health care that doesn't bankrupt the people who need it, and the many other faces of the effort toward a more livable *present.*

These newcomers were full of energy and ideas—and didn't stop to ask, or to hear, whether those ideas had already been tried. They were full of fear, anger, and urgency—and unwilling to take direction from people who'd already learned to live with, move through, or channel those emotions. If your losses and fights have been ongoing, it can be incredibly frustrating and alienating when people approach you talking only about the losses of the future. On the other hand, if climate realities have mainly come home to you in terms of the future (what's *going* to happen, what you *might* lose) you may not have reckoned with what people more structurally burdened than you have already lost—but also with what they've learned about how to work together to meet a threat.

There is an overlap, and you can find it, among what is useful, what you can personally do, and where you can belong. Sometimes it takes a little while to find, and what seems best may change as you learn more. The Climate Anxiety Counseling booth was useful as far as it went: it let me, and everyone who spoke with me, briefly practice being together in our grief, fear, and frustration. Each interaction offered a little more understanding of how climate change interplays with people's other stresses, dreams, and needs—I gained this, and at least some of the people who talked with me did, too. And the booth's openness and promise of anonymity let people speak freely and safely to a stranger. But what kept us safe, as speaker and listener, also kept us separate: not unable but less likely to figure out what we might do together.

The booth was shaped by my skills (as a listener and writer, mainly) as well as by my privileges, like the time to sit out for hours, the relative safety that being white and cis provided, and my physical ability. It was also shaped by my many years as someone whose sense of self came largely from my creative work—and limited by the myth of "individual action" that we examined in Chapter 1, that feeling of needing to do something *myself*, something new, something now. And when I first started going beyond the climate booth, a couple of the groups I worked with seemed unwilling to be organized by people whose knowledge of environmental justice came from surviving without it. If you're where I was, part of your next task is to explore your community's options for climate action more thoroughly. This exercise will help.

Mapping Your Knowledge

You can gather information from what activist or community groups say about themselves, what they say about each other, and any news about them (depending on the source, you may have to sift out some nonsense). You don't need to embrace an organization's values or practices wholeheartedly or unquestioningly to attend their events or meetings, learn more about their internal and public-facing methods, and observe their impact on your community.

QUESTIONS

Which organizations or groups near you are already working for environmental justice, emissions mitigation, or land/water stewardship and reparations? (You can add the Land Back efforts you researched in Chapter 3 to this list.)

What are their stated goals and values?

How do they align their work with the work of living systems? (Water, land, plants, animals...)

Do they also work with other groups of humans? Who and how?

Who decides what the organization's work will be, and how they'll go about it?

Who and what people, groups, places, or forces are strengthened by their work?

PRACTICE

Possibly with the time you cleared in the practice for "Zeroing In" (see page 125), do the research that will help you answer these questions. If you're a spreadsheet person, a spreadsheet might help. You could also draw a map similar to the practice for "Knowing your Neighbors" beginning on page 55, showing groups, goals, and activities, and drawing lines to show relationships and shared goals.

Existing groups—action-focused or not—may already do this periodically, but it's a good opportunity to take stock of your values, methods, and interrelations with the rest of the map!

You want to notice who's present and who's not: an all-white group in an area where people of multiple races live and work may miss important things. Observe how members talk and work together: if the balance of work is fair, expectations are clear, and people are mutually respectful as long as respect is due, that's a good start. Be attentive to the scope of the group's activity: what they do, but also, again, whom or what they leave out, and whether that's a strategic choice (for example, they're focusing on climate risks as they intersect with housing needs, rather than taking a bigger-picture approach) or a systemic oversight (for example, they routinely schedule public events that people who use wheelchairs, walkers, or canes can't attend). While white supremacist habits of mind have distorted the consciousness of nearly everyone living in the United States—along with classism, ableism, and plenty of other hierarchies—you want a group whose members can catch themselves and each other when they're acting out those habits, take a step back, and course correct.

You may not be able to find all of these in one place, so pay attention to what you can live with and what would make your contributions impossible or impractical. Collaborations across inequality can, without care, cement or even intensify the systems that ultimately hurt and threaten all of us. They can also help us to weaken those systems and build something better. Often, elements of both of those things happen at the same time.

THE STORY OF A TOWN IN THE FOREST

In the summer of 2016, my friend Mark and I carpooled from Providence to Burrillville, Rhode Island, a small town in the north of the state, to join ongoing opposition to a fossil fuel and deforestation project. Burrillville, ancestral Nipmuc land, is encircled and interlaced by woods that Invenergy, a company from Chicago, wanted to cut down to build a methane-fueled electricity plant. Several local groups—one focused on conservation, one more focused on public health and safety, and so on—joined together to fight the project. They invited the help of the FANG Collective, which builds, supports, and escalates opposition to the methane industry, and enlisted a lawyer from the Conservation Law Foundation to block the plant by means of laws and regulations. Mark and I were joining a march to draw more attention to the plant, and help people opposing it feel energized and connected. The Pots and Pans Kitchen Collective had food waiting for us at our destination: chili and cornbread, I think. People who wanted trees to live and humans to breathe were all around us.

Some Burrillville residents also spent hours at the computer researching the damage the plant would do if it were built, both regionally and to the climate, and looking up state requirements for permits. Others spent their hours in auditoriums and offices, testifying about that damage. I had a pile of postcards at the counseling booth, stamped and addressed to the state agencies that decide how much pollution of the air, land, and water is acceptable; I asked everyone who stopped that summer to fill one out, asking the agencies and the people who worked there to withhold their permission. Some people cooked for meetings, and some people knocked on doors to share information. Some designed signs and stickers to put up around the state, or donated the costs of printing; some, like Mark, called attention to the threat of the plant with the writing and art they made. Some prepared for nonviolent direct action: blocking destructive activities, or blocking ordinary activities to make a point about destructive ones, by means that don't hurt anyone physically (except, sometimes, the people doing them). In Burrillville, that might mean blockades and lockdowns at worksites, or at banks and offices that funded the project, for which the state could arrest and jail the protesters. Some people weren't committing to direct action because they had a police record. Some people's joint pain or social anxiety meant they couldn't march. Everyone opposing the plant invited people into the fight and directed them toward ways that they could help, and that were possible for them.

Power plant company Invenergy had applied to the state of Rhode Island for permits in 2015, claiming that the overall energy demand of New England would rise to require all the electricity they wanted to sell. They offered to buy the water to run the plant from the Narragansett Tribe, in the southern part of the state, for money the Tribe needed badly. They appealed to trade unions, holding out the promise of construction, plumbing, electrical, and other jobs. They made claims to legislators that the plant would be good for the state's economy.

The people of Burrillville and their allies fought the plant for nearly five years on as many fronts as they could, and were prepared to fight it further. And it turned out that the very length of the fight—their refusal to back down, the delay they created—was a factor in stopping the plant. In their final decision in 2019, the Rhode Island Energy Facility Siting Board explained that in the years since the application, New England had not demonstrated an increased demand for electricity. Residents of the region were actually using a little less energy than before, because of more efficient appliances and buildings, and utility companies were making a bigger portion of it through solar and wind. Since part of what Invenergy had to prove to the Board was the need for their project, and the region had already shown that there was no need for it, they could not build the plant.

Burrillville's victory shows us another way to understand and relate the actions we call individual and the actions we call collective. The constant challenges that slowed the project down had taken effect. People's decisions to insulate their homes better, join a community solar project, or issue the permit for a wind farm had also helped to make this difference. So had state policies and funds that helped to enable and pay for those decisions, and the people who'd agreed or voted to put them in place. I was in that Energy Facility Siting Board meeting with a sign I'd made: NO PLACE SAFE FOR FOSSIL FUELS. My palms, holding it, sweated with relief as the people around me started to cheer.

Invenergy's loss was a gain for almost everyone. One less methane plant equals that much less greenhouse gas in the atmosphere equals that much less warming equals a slightly safer world. It was also a gain for the humans of Burrillville, who would have suffered directly from the pollutants in its emissions and from the destruction of shade-providing, air-cleaning, soil-stabilizing forest. It was a gain for that forest and all the lives and relationships that compose it, from the biggest trees to the smallest soil bacteria.

And for the humans of Burrillville, in that meeting room and outside of it, working together was an individual action *and* a collective one. To care for the town in the forest, and the air and water beyond it, they each had to decide to change elements of the ways that they lived their lives and spent their time, and follow through on that decision. This is especially true for residents of the town who hadn't been organized in this way before. They let some things in their lives go, trimmed others back. They gained new skills and found new courage. They did this in the face of opposition, often from their own neighbors as well as from the company.

They also had help from people who'd already committed, in various ways, to active and organized opposition to environmental damage. This included legal support and expert testimony; it also included the strategic and tactical support of the FANG Collective.

THE STORY OF AN ORGANIZATION IN THE LAND

FANG began after Sherrie Anne Andre and a few other Rhode Island organizers returned from a Pennsylvania gathering of people opposing fossil fuel projects all along the east coast of the United States. Fighting fracking (the mining of methane gas) in Pennsylvania is rare: the companies are big employers in areas where other work is scarce, and their leases have enabled some people to hang on to family land that they would otherwise have had to sell. Areas like this, or like Cancer Alley in the Southeastern United States, are often described as *sacrifice zones*: places where ecology, including human well-being, is sacrificed for power and profit. Often, as is the case in Pennsylvania, the humans who live in sacrifice zones are cornered into working for or otherwise enabling the companies that are hurting them and their land, water, and air.

"It's so overwhelmingly prevalent that if you fight it, you lose so much," Sherrie said. "We realized after lots of crying that we didn't just want to work with each other on the basis of, 'We're fighting these things.'" The gathering also brought up an interwoven question for Sherrie and their fellow fighters: "How do we provide emotional support for each other when we're feeling so alone?"

Back in Rhode Island, as the newly formed FANG Collective (they started as Fighting Against Natural Gas, but later switched to the acronym) gathered members and understanding, they became for a while a research and grant-writing organization. They figured out what laws and regulations could be used to block a pipeline or a fuel compressor, and which explanations of their activities would get water and land defenders money: for transportation, for supplies, for lawyers, for bail. This preparation is part of why the steps that Burrillville residents took were effective: they had guidance in directing multiple kinds of attention and pressure. Over time, FANG also began to go where they were called, or offer their services—as grant writers and researchers, and also as direct action strategists, planners, trainers, and medics—to groups fighting mining, fracking, and burning.

FANG's members didn't know all this going in—they learned it. You can learn it, too, and you can learn from people who've done it about how to complement their work with your own capabilities. As they went along, they also developed practices to help the group function *as* a group. "When we would come up with a list of roles we needed for an action, or a march, we would also have these community care roles," Sherrie recalled. "Their job was to do childcare, or make sure our dogs got walked, before we had children. And also just like, make sure that there's snacks in the office, and that people are drinking water! If you see people sitting for too long, asking, 'Should we go for a walk outside?'"

Bringing the sunscreen and remembering quarters for the meters; being the person who orders the bulk beans and cornmeal, who may or may not be the same as the person who cooks the chili and cornbread, or dishes it out; planning the route to the fuel truck blockade in the next state and calculating the budget for gas and tolls, even if you can't go yourself. You may already be the person who does this kind of practical coordination and care in other contexts! If so, the change that's open to you in this moment is not in who you are or even what you do, but where you do it, and with whom. These transformations of your schedule and your habits are personal changes that add up to collective effort. And it works the other way, too: collective power is also your personal power. Everyone who helps to build it partakes of it.

Here's an exercise to remind you of that: to help you access the joy in unison. Thinking of and doing the practice part regularly, as a ritual, will probably offer you the most.

Feeling Shared Strength

This exercise is adapted from one that Laura Brown-Lavoie developed and shared. Like "Steadying Your Nerves," it works on your body through the feeling of making a sound or motion, as well as connecting you with the other beings around you.

QUESTIONS

When is it clear to you that you're not alone in your efforts to survive?

When is it hard to remember?

PRACTICE

If you're doing this solo and can hear, find some living activity that makes a noise—flowing or falling water, leaves or rock formations that pick up the wind, an area where birds or bugs or human teenagers chatter. (If you're not able to go someplace where this happens, find a recording.) Make the sound, too, matching it as closely as you can for a minute or so, or just open your attention to the sound. If you're D/deaf or hard of hearing, attend to some other aspect of the scene, like light or motion, and stay with it for a minute or so—moving in response to it if you like, or tracking it with your eyes. The general goal is to align and attune yourself with some aspect of the activity you're paying attention to, in a way that you can do and that lets you feel connected with it.

If you're doing this as part of a group, you can match or tune in to each other's sounds or motions. Building from quiet to loud and back to quiet sound, or from small to big and back to small gesture, is a good and dynamic variation that helps you attend to and connect with one another as well as your surroundings.

KINDS OF POWER

Not long after the march in the forest, my friend Mark left Rhode Island on a longer walk: a barefoot pilgrimage across the continent to bring more attention to climate change and raise money for the FANG Collective's work throughout the country. "We were going to yell at the streets until they grew ears," he wrote in a blog post after a protest he joined in Jacksonville. "Yell for clean water until all the streets no longer have cars." On his walk, he joined water protectors' camps and actions, both bought and salvaged food, was stopped by police and offered cash by strangers, and recorded and wrote blog posts and poems.

If you consider the Climate Anxiety Counseling booth and Mark's walk side by side, you can tell that both of us started out as creative writers, people in the habit of making symbols and gestures matter. You can tell how each of us played to our strengths—his willingness to embrace physical discomfort, my openness to conversations with strangers—as well as the ways that being white and nondisabled reduced some of our risks. You can also tell that although Mark was taking a lonely road, he was supported all along that road by collective care, and he used power particular to him to honor, continue, and equip collective action.

The kinds of power we bring to fighting and caring for the living world can help us out; they can also trip us up. I wrote earlier that Invenergy's loss was a gain for almost everyone, including most of the parties involved with the campaign. But it also left some people out—and in doing so, missed a major opportunity to set right a historical wrong, and to build more just and lasting power to care for each other and life on Earth.

Sherrie Anne Andre recalls the frustrations of negotiating with the Narragansett Tribe about their offer to sell water to Invenergy before the state denied the siting permit. Without an understanding of oppressive histories and systems, Sherrie says, it's hard to see why a community might make a deal that would hurt them in the long term. But if that community has been and continues to be an oppressed one, the trade-offs they're willing to make to meet community needs can be complicated. The Narragansett Tribe's ancestral land and water, which covers much of the colonially imposed state of Rhode Island, is a sacrifice zone in the sense that they have little say in what happens there, and that much of what happens there harms them. That short-term benefit—the payment to the Tribe for the sale of the water—could, if it was delivered and distributed with care, have improved the material circumstances of a lot of Narragansett people, even as the plant was polluting the land, air, and water that they need, too. It's the same calculation that came up in the last chapter when you were contemplating degrowth, or leaving your environmentally destructive job, scaled up to the level of a community and intensified by years of induced, genocidal scarcity and suffering.

In his letter requesting termination of the water deal, which is part of the public record of the application to build the plant, a tribal official says that opposition to the deal was coming from "a small group of persons [who] have wrongfully claimed that they speak for the Tribe," that "the turmoil was exacerbated by individuals outside the Tribe," and that these groups together "have an agenda unrelated to the best interests of the Tribe and its clear economic needs." An Invenergy employee wrote back, in part, "We entered into the [Water Sale Agreement] not only as a means to secure an additional possible source but also to assist the [Narragansett Indian Tribe] in some economic manner, even if we didn't use the NIT's water."

For the record, I don't think Invenergy genuinely had the Tribe's best interests in mind as part of their business plan. But I tell this story because you're going to encounter situations like it as you become more involved with climate and environmental organizing. When people or groups who seem (to an outsider) like they ought to work together for the living world, or a piece of it, are instead at odds, it's worth trying to sort out why. Often, it's a question of needs that seem to be in conflict—*either* you get the money *or* you get the slightly less polluted air, land, and water; *either* you get the victory *or* you get the goodwill. But when you notice that conflict, it's time to ask more questions: Is the incompatibility real, or is it the result of too narrow or biased a view of what's possible, whose knowledge is relevant, and whose lives are worth living?

You're more likely to overlook the needs and expertise of people you've been taught don't matter—even if those people are you, or like you. And that in turn can lead you to choose climate and environmental goals that are harmful in their incompleteness. If the same people who are always being told to suck it up are once again being told to suck it up, that's not a just process. And it's not a sustainable one, either, because it treats humans—once again—like resources to be extracted, and passes that extraction down the hierarchies of power.

This is true both within and among groups. Radical care community of practice Climate Critical, in their research on burnout in the climate movement, learned that "Black and multi-race identified respondents often cited systemic and structural racism as causing and exacerbating their burnout. Many participants stated the challenges of working in performative, white-dominated spaces where their voices were tokenized and their roles burdened them with demands of representing communities on the front lines of climate change." In other words: if you're a marginalized person (and this applies especially to Black people, Indigenous people, and people of color) working against environmental violence, these systems have already hurt you, so you do have insight into how they work and how they need to change. But in a group that's not being *extremely* purposeful and scrupulous, people who've benefited more from these systems and suffered less are going to try to squeeze that insight out of you, along with a lot of work, and maybe take the credit, possibly while also making you feel outnumbered and lonely on the day-to-day.

Or you may—even without conscious intent—be in the position of making the demands and taking the credit. This is likelier if you're white and coming from a privileged background, but you can fall for it even if you're not: the thing about extractive systems is that they're very good at getting you to keep them going. They are big enough to trap all of us. And yet it is out of their pieces—which include ourselves, shaped by them as we have been—that we have to build a more livable world.

THE LONG HAUL

I was an outsider to the Burrillville campaign—not a resident of the town, not a member of the Tribe, not a lawyer or an organizing expert. But even from that limited perspective, I can see that in the case of the water sale, the people fighting the power plant could have made more commitments to find another source of income for the Narragansett Tribe, so that *only* Invenergy would lose out. I can see how that would have been more fair in itself, as well as more strategic, building the narrower zeal to protect the land from an imminent threat into the possibility for a longer haul, a broader view of what threatens the land and how to protect it, a more collective effort to survive.

And you know what? The town could still make things right—maybe with money, or by weighing in on the Tribe's side the next time they have to deal with the settler government, or in some other way that would reveal itself if the people who'd fought the plant also fought for this reparation. Town officials and tribal officials could pick up that process tomorrow. I referred to the Burrillville "victory" earlier, but there are no final wins or losses on Earth, no closed chapters. History teaches us that while we're alive, there's always the next morning.

So it's worth it to take care, and often different *kinds* of care depending on your own position and power, to make sure that the power you're building is fairly redistributed among the people who are building it. Climate Critical's recommendations include massive changes for formal environmental groups (especially large organizations) and donors, up to and including closing their doors if they cannot operate without exploitation, and attempting to repair damage they've already done. If you're a relative newcomer to active climate efforts, or other organizing or community efforts, the results of the questions and practice for "Mapping Your Knowledge" earlier in this chapter can guide you in joining or supporting a group whose day-to-day workings and relationships are more often equitable and not extractive. You can also be alert to when a group is creeping in an extractive or tokenizing direction—or when you're playing into that dynamic yourself—and insist on pausing and walking it back. And you can borrow the Climate Anxiety Counseling lessons of listening and asking questions before you make statements, and of asking, "What have you tried?" and "Is there a reason you haven't tried X?" before making suggestions. There are more exercises in the next chapter about productive conflict and confrontation, how they might get you feeling, and what you might do within them.

Once you've reached the point of getting involved with a collective response to climate change (Clearing the time! Making the phone calls! Looking up the meeting dates! Giving the rides! Building the barricades! Clearing the hours again!), it can feel infuriating or exhausting to walk back a plan, be more scrupulous, make provisions for more people's needs—especially since the forces driving climate change have no scruples whatsoever, and are going full steam ahead. But Citizen Potawatomi scholar and teacher Kyle Powys Whyte has written and spoken about the ways that urgency in climate and even environmental justice movements can—strange as it seems—lead to greater losses through carelessness, band-aid solutions, and reinforcing histories of inequality. And the Climate Critical report on burnout shows how reliant climate and environmental organizing can be on people going beyond what they can really give. This is another reason why the questions and practice for "Zeroing In" are useful: you can start to find your place in this work now *and* move into it deliberately, remaining willing to revisit your course, your relationships, and your vision, whether a demand to go beyond your abilities is coming from your own sense of urgency or someone else's. Here's an exercise to move you through that sense of urgency—which may feel like anger, fear, irritation, helplessness, or something else.

Lifting and Letting Go

The practice here can help you to be guided, rather than provoked, by what you feel and know about the world's changes and losses. You can do it alongside any course of climate or community action that you're considering or embarked on.

QUESTION

What about climate change, environmental injustice, or ecological loss is heavy on your mind today, or feeling like it demands urgent action?

PRACTICE

As you answer that question, pay attention to how your mind and body are feeling. Enter that feeling fully, and then let it fade (like a wind that picks up and drops, or the feeling of cars approaching and passing as you walk beside a road). To make this more of a practice, deliberately bring your mind back to that same change or loss or injustice—that climate reality—each day for a week, letting the memory and feeling intensify and subside.

If you're doing this as part of a group, you can do it when you're physically or virtually together, or you can call or message each other for support and recognition.

It may help to remember that doing differently, shifting our efforts and our loyalties to the livable and living world, doesn't necessarily mean doing more. A lot of us *can't* do more in the sense of "on top of everything we're doing now." But doing differently can feel like doing more because it isn't a habit yet, and because it often means coming up against structures that are both familiar and bigger than you. And then, when you manage to make that shift, you have to *keep* making it. You have to keep negotiating, keep checking, keep dealing with other people.

Because climate change places us under urgent pressures, because any ordinary thing that one person can do isn't enough, and because working with other people is slow and hard, it can be appealing to commit to a response that is individual and extreme: a grand, symbolic gesture. One of the things I appreciate about Mark's barefoot walk is its sharp difference from ordinary life, and his willingness to be transformed by what he knew and felt about the world and the ways it's changing. Within the systems that trap and limit us, he found a way to move according to his own inclination while also doing many of the connected, collective things recommended in this chapter: raising money, joining protests, strengthening principles. While I never asked him about it, my guess is that doing those things without some powerful and unusual gesture didn't *feel* like enough, in the same way that just being a writer hadn't felt like enough for him, and in the same way that the booth stopped feeling like enough for me. No one, alone, is enough for the extremity of what we're facing—but we're also not facing it alone.

There is a place for more extreme, dramatic tactics—and not just symbolic ones—in this effort to transform the world. You may even find that it's your place. But all our actions have the chance to matter. And even when there's a missed opportunity to do the right thing within the larger, ongoing effort to meet climate change, like the breakdown between the Burrillville campaigners and the Narragansett Tribe, ongoing effort will bring you more opportunities to make different, fairer choices and build better, more mutual relationships. There are few certainties within our climate present and future, but that's one of them: the chances to deal justly with and care for one another will keep coming.

CHAPTER 7

FIGHTING ON OUR OWN GROUND

FROM ANGER (AND FEAR) TO CONSTRUCTIVE CONFLICT

YOU ARE HERE

This chapter invites you to learn these things:

- Climate rage can be a useful tool, especially when it's shared.
- Using conflict intentionally and positively helps groups of people working toward a shared goal improve honesty, calibrate fairness, and choose their strategy.
- Some of the most powerful ways to fight for environmental justice are rooted in the places you live and the people you love.

DON'T BE AFRAID (TO) BE ANGRY

Gina Rodríguez and her husband Julian were visiting her parents in Johnston, Rhode Island, when Julian idly picked up a copy of the local newspaper. "And he said, 'What is this?' Gina recalled in our interview. "'Cause it had, on the front page, a picture of some white guy in front of the natural gas tank. And the headline was something like, 'National Grid to build a liquefaction facility.'"

"Are you *fucking* kidding me," she remembers saying. "What does that mean?"

Julian slapped the paper down on the table. "This is it. This is it," he said. Thinking back to his anger now, she laughs. "I thought, 'Oh God, I know that face. Here we go.' We instantly knew: 'No, this can't happen, we have to stop it. We have to do everything that we can to stop it.' While also in the back of my head knowing it would be very unlikely that we could actually stop it. But maybe stopping it all the way wouldn't ever be the goal? We went into it knowing, 'We might never stop it, but we're gonna make them lose every dollar that we can, and we're gonna hold them back, and we're gonna get some energy behind us and we're gonna change the conversation. Even if we don't win.'"

Gina and Julian were able to quickly direct their rage at a source of climate injustice, and use that rage to drive their fight—a fight that I and many others joined. Thinking in terms of systems and being alert to our power within them helps us to do this, rather than "punching down"—taking our anger and frustration out on people who can hurt us less. In this case, a fossil-fueled energy company planning an expansion in a low-income neighborhood was an obvious target. But it isn't always so easy to know where our anger should go, or how far we should follow it. Much later in the campaign against the liquefied natural gas (LNG) plant, my husband, James, and I stood on a bridge across the Providence River, upstream from the proposed plant site, angry at each other. He wanted me to say I wouldn't join the group in nonviolent direct action, things that might slow down the plant further but could lead to my arrest or imprisonment. I wouldn't promise. We stared at the water, unable to connect.

Well-informed, well-grounded anger is a weapon for climate and environmental justice. Complicated anger, between people who share a desire for all of us to survive, is painful and can stall or derail our efforts and divide our loyalties. It can also lead us to better strategy, greater honesty, and more ways to act on love.

This is a moment to look for *your* climate rage. The goal isn't to stop being angry, although approaching any emotion on purpose can help you get better at releasing it on purpose. If your anger is justified, though, there is no need to stop it. Rather, this next exercise can make you more familiar with where your anger comes from and where it ends up. This, in turn, can help your decisions about where your anger leads you to be more informed, more purposeful, and braver.

Directing Your Climate Rage

Anger feels powerful to some of us. For others, it's frightening to feel and we try to avoid it. Some of us—especially if we're in positions of relative power—are happy to dish it out, but we try to squelch it in others. What we tend to do with our anger comes from our histories: *Is it safe? Can I get away with it? What will it cost me?* And even: *Could it lead to something better?*

QUESTIONS

If you get angry about climate change, when do you get angry—what things are happening around you or within you when your anger rises? (You can name many examples, or just go with the first thing that comes to mind.)

What could your climate rage empower you to do?

PRACTICE

As or after you reflect on these questions, find a position that you can hold for a few minutes with small adjustments. Close or unfocus your eyes, and pay attention to your breathing in and out. Note your sensations, starting with the places where your body is supported by something (the ground or floor, your chair or bed) and moving toward the top of your head. You're feeling for the places in your body where anger collects, and what it feels like there.

Optional: Imagine your anger as light or heat radiating from that place or those places. Follow the direction of the beams with your mind's senses. Notice if they're just radiating around or if they're pointing at something or someone. Now pause and check if you want to direct your anger differently. If you do, picture that—both the change of direction and where the light or heat hits. Check how you feel. Finally, consciously picture the light or heat fading (you can always get it back), but make a note of where it landed.

If you're doing this as part of a group, decide before you do the exercise whether you want to share what you observed and/or check in with each other if you're planning to do the practice routinely for a number of days.

To our collective efforts to heal the living world, we bring everything about that world: true and untrue, useful and harmful. We bring our skills and our love; our material situations; our personal, family, and cultural histories. And our ability to work together effectively depends on understanding those forces, how they shape our reactions, and where they might send us.

"DAMN! NOW THIS?"

When I think of 2017 in Providence, I remember women and microphones. The microphones are in big pale rooms or on the steps of big pale buildings, sites of the state's power. The women—two women in particular, Gina who you've met, and Monica Huertas—have come to these sites to defend their neighborhood, the Southside of Providence. They speak strongly, holding themselves upright or driven forward by the force of their anger and their love.

When Monica and her husband Juan bought the house on the Southside, a few blocks up from the Port of Providence, she felt like she'd finally made it: from homelessness and raising her kids with a partner in prison, to putting herself through school for social work, to actually owning a home together. While Juan worked as a carpenter, Monica was parenting their kids and organizing against police brutality with Direct Action for Rights and Equality (DARE). Maybe they were going to be okay.

Monica thinks now it was through fellow DARE members that she learned of the first public meeting about the LNG plant. Liquefying natural (methane) gas lets companies like National Grid store much more of it, meaning that if it explodes, the explosion will be bigger and destroy more. Set between the Port and the highway, the plant would be in destructive range of other fuel and chemical tanks and pipes: a leak or explosion at one could ignite others, releasing poisonous and flammable fumes. Explosions at similar plants have required evacuations for two miles around; one in Algeria killed twenty-seven people, injured seventy-four, and did $800 million worth of damage. And the operation of the plant, by increasing the amount of heat-trapping gases burned, would put its structure at greater risk: warming the climate further would increase the likelihood that a storm surge could flood the site and trigger a disaster.

Monica and Juan's house was roughly a mile from the site. It was also, she learned, built on soil that had absorbed years of industrial pollution—so was the whole neighborhood. Digging to build the plant would release those metals and other compounds into the air, and their third child, Alex, was already suffering asthma attacks that regularly took the family to the emergency room. An explosion at the plant could take out the children's hospital, too.

"Jahshua was two months old when we bought this house, so [it started] right around that same time," Monica recalled. Jahshua, her youngest (age four during this interview), tried to grab my recorder for the fifth time, and we shooed him out. "I had gone through all these things—now we're able to buy a house, and it's like, 'Damn! Now this? My house is polluted, and it's sitting on toxic soil.'" And it was about to be downwind of yet another project that would heighten the risk to her family and her neighborhood.

Neighborhoods like Monica's and Gina's are another example of *sacrifice zones*, areas where industry treats life as expendable. They can be harder places to care for or fight for because they don't fit the idea of "nature" or "the environment" that many people have in their minds: landscapes (like Burrillville's) where more-than-human beings outnumber humans and human-built structures—as if water and air weren't shared and didn't travel—and because of who's driven to live there. Humans who live in sacrifice zones are usually people culturally devalued by their city or nation: people of color and poorer people (who are often also queer and trans, older, and/or disabled) are driven there by higher prices elsewhere, or have industries piled into a place where they're already living. These are people who often have strong relationships and great ingenuity—all of which they'd much prefer to use for things other than survival. They'll also have a harder time surviving or rebuilding if acute disaster strikes, and often can't afford to "just move." (Remember the damage from Superstorm Sandy in Chapter 1, still unhealed in Red Hook years later.) That's the Southside of Providence: an asthma hotspot, a redlined district, a home.

Gina Rodríguez also still lives in her house on the Southside, and I talked with her there. "My dad fought in the Cuban Revolution," she told me, when I asked why she was so prepared to direct her anger when the opportunity came. "My whole life, this idea of fighting for something that you believe in and being able to win was very real. I think I was just brought up to be an activist. Without having a framework for it—my parents didn't take me to protests, I started that in my teenage years on my own, but the idea that you fight something that's unjust, that was part of the air." She laughed, and hopped up to check a pot on the stove: her kids, seven and four, were on their way home from capoeira class.

The work of the house moved in and out of Gina's stories. On her first date with Julian in college, they ended up talking all night about Palestine and climate change. After they graduated, they started Confluence, an annual "intentional gathering" of activist friends from all over the Northeast. When they thought of buying a house together someday, they imagined installing composting toilets and graywater systems. "And then fast-forward a few years with kids and we now own a house . . . and the reality of money and time and childrearing kicks in. And then I'm also thinking, 'Well, fuck! If National Grid is gonna build this liquefaction facility, my little graywater system or my natural soap is not gonna make or break anything'"—they wouldn't be the tipping point for climate stability or environmental risk. The house where the kids, now home, were eating mac and cheese and showing off their capoeira moves is across a major street and a few blocks down from Monica's family's house. Like theirs, it's well within the evacuation area of the LNG plant.

As Monica learned more about the plant proposal, she began to both talk with Southside neighbors and forge ties with other organizations: No LNG in PVD, the coalition called themselves. People started to gather around the campaign, and one of those people was me. Julian stopped by the Climate Anxiety Counseling booth in the summer of 2015 to invite me to an information session.

The Southside wasn't my home, but it became my fight. The abstraction of "collective action" now had hands and arms and legs and voices. If I added mine, I'd be joining a chorus. If I stepped out my door, I'd be walking with them. That was the beginning, for me, of two years of meetings and hearings, research sessions and door knocking, goofing at the bar after a day painting banners, and spooning out rice and gandules at neighborhood Earth Day events. I brought Monica and Alex dinner when Alex was hospitalized with another asthma attack. We handed out flyers with the governor's number on them at the stoplight closest to the plant site, and held homemade signs and banners in front of the tanks and pipes that National Grid already had in place: HIGHEST ASTHMA RATE. THIS IS ENVIRONMENTAL RACISM. STOP POISONING OUR STATE, OUR COAST, OUR PEOPLE.

Fighting climate change can feel like punching the air. Yet every day, it's hurting you—you personally. At the Climate Anxiety Counseling booth and elsewhere, people have for years been telling me how hard it is to connect that personal hurt and anger with a course of action big enough to meet whole governments, economies, and their interactions with giant elemental systems of land, water, and air. But when that opportunity for connection is presented to them—to us—we often back off. It doesn't seem like the right size for our lives, or the kind of thing we'd do; we assume it's incompatible with our present survival, even though our future survival depends on it.

It's worth releasing those assumptions at least long enough to test them. And it's in that testing process, which you explored in the previous chapter—What *can* I offer? What *will* I do? How will I know if I'm in the right place?—that we learn how to direct our energies (including our anger) toward building a life that makes sense and that we can keep living.

FINDING THE FORM AND TERRAIN OF YOUR FIGHT

When you think of fighting climate change, do you think of people locking down to construction equipment on a pipeline route? Sit-ins in government offices, as congressional staffers did in 2022 to push the Inflation Reduction Act back into play? Do you think of governance, policy, law, regulation? Coordinating interpretation, wheelchair transportation, and portable toilets for a rally? Watching people's babies? Planting trees or grasses? *Being* a tree, a colony of soil bacteria, or an ocean? Not all forms of fighting are available for all forms of life; all fights, like all communities, require multiple sets of skills and strengths, acting in concert, and choosing the methods that they are actually able and willing to carry through.

While I was involved, No LNG in PVD did a lot of our fighting with paperwork. We met at Monica's, usually. Gina and Julian's kids played with Jerimy, Victoria, and Alex (and later, Jahshua). Andrew, an environmental engineer, brought his daughter (later, both daughters), along with his familiarity with state regulations and technical language. Aaron kept tabs on meetings, hearings, and requirements for public engagement; David brought experience from organizing for safe homes and clean water; members of the FANG Collective shared their years of knowledge from fighting fossil fuel projects throughout the country. We gathered in the midst of the rest of our lives to eat pizza, hammer out strategy, and divide up tasks.

"Eeeeeeeverything that I learned was on the fly," Monica said. "The grant-writing stuff, the technical things about the history of the plant, and also the science around fracking and liquefied gas. I'm by no means an expert in these things, but after all these years of doing this I feel like I am an expert in my community and in the things that we need." As the fight progressed, she began to raise her voice publicly, to say with her words and her tone: we're not *asking*, we're claiming what we should have, a livable life for ourselves and our kids. These were also years of learning how environmental racism tied in with the other kinds that she'd lived through: "I knew we were always getting discriminated against, I just didn't realize that it was also with toxins and trucks and the trains."

For my part, working with No LNG in PVD made me face the cost of my relative safety as a white person living in a neighborhood that wasn't a sacrifice zone (yet). I learned when I'd be contributing more by listening rather than talking. I also learned the names of state and federal agencies I hadn't known existed. I transferred longtime skills as a reader, writer, and teacher from composing poems to editing press releases, translating technical language, and summarizing talking points for public testimony. Members of the public *can* demand environmental testing and impact assessment for a fossil fuel project, as well as opportunities to state in writing and in person why they don't want it, but you can't demand what you don't know about. So No LNG in PVD started making those demands, and letting other Southside residents know that they could, too.

With Monica's leadership, we used those demands to put off the day when construction could begin—as with Burrillville, each hearing that had to be scheduled bought us a little more time.

But I grew more and more furious as people from state environmental agencies, who I'd have thought would be on the side of life, said they couldn't consider climate change—its causes *or* its effects—as a reason to deny National Grid any permits. They talked about the plant as though it were going to sit quietly in the floodplain and do nothing, as if its operation wouldn't be *expanding* the floodplain, polluting the air in the neighborhood's lungs, trapping more and more heat near the surface of the Earth.

We went to the hearings anyway, bringing snacks for the kids; we testified for hours, residents and engineers and farmers and doctors and students and even a couple of senators. At the second hearing, the state employees were looking bored, leaning on their hands or marking time with unfocused eyes. And then Gina Rodríguez took the microphone.

"I want to speak to you," she said, addressing the agency's director, "because two weeks ago I watched you mention the provisions for what you can regulate and what you can't. And I can see it in your eyes, I can see your humanity and I can see that you want to do more, and that you wish you could do more. I first have to tell you that I felt angry at your response, I felt enraged. It turned into feeling discouraged, and now it's settled in disappointment." Her voice rose, intensified. "I am disappointed in each and every one of you, because as a mother I cannot believe that whoever raised you raised you to be a coward, and we cannot afford for you to make a cowardly decision—"

I was listening but not looking at her, so I didn't notice the motion around her until other people waiting to testify rose to their feet. They were trying to get between Gina and the state cops and building security who were closing in on her. I stood and joined the ring of people around her, and others did, too. The moment tightened, then passed: she finished speaking, the officers stood down. When the hearing ended, we all walked her to her car in the too-warm November night.

"That was more of a performance than it was commenting," Gina told me. "Because I knew: we've all heard the science. They don't listen to the science. And sometimes that role, the role of the artist-activist in the room, is to hit an emotional chord. Calling the head of [a state environmental agency] a coward, saying his mother did not raise him to be a coward, was not about him necessarily, but it was about—you're a human being. And you're making inhumane choices." The terrain of Gina's fight was emotion. She used her experiential authority and her skill—she's also a poet and performer—to get angry on purpose, taking up space and compelling a reckoning.

What would have happened if one person from that agency, or more than one person, had risen to Gina's challenge by changing sides? Had said to the cops and to the colleagues who'd summoned them, "Let her finish"? Would what she said have changed the outcome? Not by itself. But would *what they did* have changed the outcome? I don't know. I think it's possible. But no one took that leap. There may have been people at that agency who, within their own offices, argue for stopping environmentally destructive projects. But in the end, however they felt as people, as an agency they aligned themselves with the forces of exploitation and death. If they felt climate rage, they let its potential run out.

Taking a leap for environmental justice is especially hard when it brings us into conflict with some of our present communities, and so with our own immediate interests. If the people Gina had addressed had publicly shifted their loyalties, there might be negative consequences next time they were up for review at work, or in relationships with their colleagues; if the agency as a whole had denied the project, eventually National Grid would probably have taken them to court, with what consequences I don't know. Putting your career on the line for an uncertain outcome can drain away your anger, leaving you in fear and helplessness. But is that where you want to live? What would make a different choice a more appealing one?

Here's an exercise to help you think about it. This exercise does *not* ask you to position yourself as what Diane Exavier calls "the Lone Ranger," taking a solitary stand. That is not a sustainable position. Instead, it asks you to enact our interdependent reality, gathering community to hold you up as you do something potentially difficult and costly, or being part of the circle that upholds another person in insisting on community well-being.

To Fight on Your Own Ground

This practice is especially potent if you do have a job or role where you're making or influencing climate-relevant or environmental justice decisions, even local ones—if you work for the water board, the zoning commission, or the health department. It's also useful if climate and environmental justice, or other forms of justice, are at stake in the activities of your workplace, or in the PTA, at your mosque or church, within your union local. This practice won't make it easy to go against your short-term interests: your financial security, your ease and convenience. (If you have more social capital—say, if your race, your senior status, or your credentials offer you some cushion—you can take advantage of that!) It may produce emotionally painful conflict with the people whose lives are entwined with yours. But it will give you some courage to face those moments, and some reasons to move through them to a state where change is more possible.

QUESTIONS

When was the last time you were faced with a chance to shift your climate loyalties, even a small one? (Taking on or refusing a client, embracing or rejecting a policy . . .)

What did you do?

What was the cost of your choice?

What *could* you have done?

What will you do next time?

What might be the consequences of that choice?

PRACTICE

Make a plan for what you'll do if and when this moment comes to you. Think what you'll say and who to say it to. How can you lay some groundwork for weathering any consequences *before* you speak up? If your work or other roles are unlikely to put you in a position to make or influence these decisions, plan what you might offer someone in that position.

This could be an offer you make as a friend: help with a résumé or CV, public outcry if they lose their job, listening and affirmation if they stay on but lose coworkers' regard or cooperation. It could also be an offer you make strategically, as part of a negotiating group or organization—"If you do X, you'll have our support for Y"—although that can bring its own difficulties.

BUILDING A PRODUCTIVE PRACTICE OF CONFLICT

I'm not going to tell you stories of anger and hurt as they show up *within* groups, among people who are trying, with goodwill, to work purposefully to meet climate change together. For one thing, you already know how those stories go. People disagree about strategies or methods or priorities, and/or someone disrespects, neglects, or fails to care for someone else. They either bring it out right then or they let it fester and it comes out later. (I'm focusing here on people who are trying, and sometimes failing, to treat each other right, rather than someone who's persistently treating others badly for their own weird purposes—but the line isn't always clear.)

All the intragroup conflicts I've contributed to or witnessed involved real feelings of injury, anger, and pride, and some real instances of unfairness, disrespect, and even exploitation. Some of them ended collaborations that had had fruitful outcomes, broke bonds that had been warm and nourishing, or dissolved groups that might have been able to do more together. But when I went to write down the stories of what actually happened—who failed whom, who responded in which ways—they were so, so incredibly boring.

That's because conflicts are rarely only about what they seem to be about. In their workbook on productive conflict, *Turning Towards Each Other*, Weyam Ghadbian and Jovida Ross propose that conflict arises when something in the present reminds us of ways we've been hurt or failed in the past—when a "core need," as they call it, feels threatened. They write, "Conflict unveils systemic traumas and the ways oppressive systems and violent people have used power in extractive ways against us across time, space and generations. When we avoid conflict or move through it carelessly, we end up acting out those structural patterns unconsciously, even if we are from an identity harmed by those systems. . . . This can end relationships, organizations and movements."

Their workbook asks people to use quieter, less active times to become familiar with their own core needs and their reactions when those needs feel threatened. Because so many of our automatic inner responses to fear and threat live in our bodies, using physical (somatic) methods can help us slow down and shape our outward responses with more purpose—whether the threat we perceive is from a present disaster, the fear of greater climate chaos, or a person we're trying to collaborate with.

Respecting Your Reactions

This exercise builds on ideas from Diane Exavier, as well as Weyam Ghadbian and Jovida Ross. Its purpose isn't to avoid conflict, but to approach it from a position of physical self-awareness, so that you and others can access conflict's benefits. The questions in this exercise are painful for some people, and unlike most of the questions in this book, I encourage you to start by answering them in silence. And if you read on and your instant reaction is "Nope," skip this one.

QUESTIONS

Where in my body am I braced for a threat or conflict—expecting it, tensing up for it?

Where do my expectations come from?

Have I felt this before? What was happening then?

What might I want to do differently this time?

PRACTICE

Create a small but uncommon gesture to make when you notice that you are braced for a threat. Practice making it when you are angry and when you are frightened.

If you're doing this with a group, settle on a shared gesture and make it a signal. (It will need to be one that everyone present is capable of doing.) Use it whenever you notice your own anger or fear; watch for it in each other, too. Once it becomes a habit, if it's helpful for you, you can also try doing it in situations where you and your group aren't the only people around, as an alert or a request for support.

This practice has things in common with a safe gesture (like a safe word), with the glance you give and receive with a friend across the room when it's time to leave the party, and with a secret handshake. It's a way of signaling to yourself, and maybe to the people you trust: let's proceed with care. It's not a replacement for fuller communication about what you need to confront. But it can remind you that when you're feeling threatened, even by each other, you have something that you share. And it can give you a way to check in with yourself about whether and how to bring your anger forward.

Anger can be a fitting response to what's happening within a group, as well as what a group is trying to fight or stop, and airing it gives people the chance to learn from it and alter their behavior. Conflict can also be a necessary tool for deciding how to fight. Yotam Marom, an organizer and group facilitator, wrote in a Medium post that conflict is an indispensable phase or stage in creating a strategy and proceeding toward a shared goal. It's how you make sure you do share a goal; it's how you make sure the methods for reaching that goal don't screw anybody over or leave anybody behind. "Strategic clarity requires conflict," he concludes, "because focus, boundaries and honesty are integral to good strategy.... None of those things are possible in a conflict avoidant group because conflict avoidant groups stay on the surface in order to protect themselves. Conflict avoidant groups appease each other, shy away from the details.... Conflict avoidant groups don't tell the truth." (I recognized these descriptions from my own behavior in groups I've been part of. It was embarrassing.) Marom adds that even when there isn't a clear solution and what's there is still tentative, raw, and uneasy, "the point is the group practicing, exercising its capacity to be in struggle, moving toward greater understanding and depth and alignment that will allow them to go into the fire together." Conflict is productive when it clarifies and strengthens what we share, and what we'll do or demand for one another.

BECOMING BRAVE FOR WHAT WE LOVE

In *Turning Towards Each Other*, Ghadbian and Ross also ask people working together to affirm what nourishes them, stabilizes them, and gives them purpose, both within and between moments of conflict: "Naming what brings us together and what matters most to us can be orienting when times are tough." What matters most to you? What will you fight for? What do you want to give life to? I wish I'd known to invite other members of No LNG in PVD to do this with me, or to recognize their invitations when they were offered, in moments of tension and indecision. I wish even more that I'd been able to reach for it with the person I'm married to.

After the regulatory and legal efforts to block the plant failed, there James and I were, standing on a bridge. The group had come to the point of choosing between letting the fight go and trying to advance it with direct action—the stuff that can get you arrested, charged, and jailed. James asked if I would do that. I said the group hadn't decided yet.

"But you can decide," he said. "You can say you're not going to do that." The bridge we were standing on crossed the Providence River where it separates downtown from the high ground of the East Side, upriver from the hurricane barrier that shields downtown from storm surge but doesn't protect the neighborhood where the plant was proposed. Where it would now be built unless we escalated. Maybe even *if* we escalated. The place where so many fuel pipes and chemical tanks already stood, waiting to be broken by extreme weather. Where my friends lived, the people I'd come to love and wanted to protect. I felt like I'd be abandoning someone no matter what I did. I had so many people who loved me, so many people to love and be loyal to, and I'd never felt more alone.

If you love someone, you want them to be whole, to be well, to live freely. Climate change makes this less likely for all humans, and so one way to act on that love is to try to limit climate change. But many of the ways to do that—including some of the most powerful—have the risk of profoundly changing, if not ending, your life with those people. If part of what you love about your life is living with someone, and jail separates the two of you, the separation as well as the harmful conditions may change the way you deal with one another. If your salary is what keeps your household going, an action that your employer might use as an excuse to fire you, or that takes you out of work for a stretch of time, is a tough trade for potential, future freedom and well-being that may not even come to pass. And this can be even more true if you've already made one or more of those trades and lost something you valued or needed without bringing about the change you *also* needed.

Eventually, the group decided not to proceed with nonviolent direct action, and my own decision was put off for a while. But James wasn't wrong that if I'd gone down that road, our life—the practicalities, but also our habitual, familiar, and chosen ways of loving one another—would have had to change, in ways we can imagine but can't be sure of. It's also true that the climate crisis and all of its accumulating, interrelating effects could change our life drastically anyway: a hurricane could tear open our house or send relatives to us in need of refuge. Scarcity or hoarding of materials or labor could cut off our access to a lifesaving treatment. The work I do for money could become irrelevant or obsolete. It can seem especially scary and dismaying to change the shapes of our lives before they're taken from us. Yet making those changes on purpose can give us a better chance of weathering them than waiting until the shapes of our lives become collateral damage. We don't choose what will happen, but we choose what we do.

I'll admit that I was relieved when No LNG in PVD decided not to take the fight to another level. I was also frustrated: I didn't know how to say to James that we were all on the same side, that by fighting the plant I *was* fighting for our life together. I was angry at him for reasons that had nothing to do with him, that were lodged in my history, about being controlled and having what I want not matter. I love our life together, and I was, and am, afraid of losing it. I was, and am, afraid of the next time I'm faced with the choice of putting that life in the balance. I am ashamed, because the plant was built, and I'll never know what would have happened if I'd lent my weight to fighting harder, had the arguments about strategy, taken the risks.

You'll hear in the next chapter about someone who made this decision differently many times, and why, and how it went for them. In the meantime, here's an exercise to do when the feelings pile up.

To Howl Together

Noise demonstrations have a long and powerful history. My city has held some to drown out Nazis, and I've lent my voice to waves of sound outside immigration detention, to let the people imprisoned there know they weren't forgotten. If you want both company and volume, you can join up with an existing group that's working for a related kind of justice—it could be the beginning (or the continuation) of a beautiful friendship.

QUESTIONS

What's an opportunity to respond to climate realities that you or your community has let pass by?

What has been lost in the wake of that decision?

PRACTICE

Find a place that can accommodate loud sound: it might be an already-loud place, a place without a lot of humans around, or a noise demonstration or disruption of some unjust activity. Choose the place with consideration for who'll be disturbed or distressed by loud noise there, and who'll be helped, encouraged, or eased. Go there and make as much noise as you possibly can, pouring all your rage and all your grief into sound.

If you're doing this with a group (highly recommended), try to hear or feel each person who is with you, within the wave of sound you're making together. If you can't go places, record yourself being loud, or another loud noise, and ask someone else in the group to bring the recording with them and play it.

Direct action isn't the only thing that works. Burrillville ousted the power plant without it, and in 2018 Rhode Island residents blocked the state's major water utility from turning over control to a corporation (something that tends to lead to lower water quality, higher rates, and shoddier repairs). Similar fights throughout multiple lands, with and without direct action, have been at least temporarily successful: residents of Cancer Alley stopped two petrochemical plants in 2022, and land and water defenders using a diversity of tactics successfully resisted the Keystone XL pipeline until the company backed down. No victory is permanent, and anything that you defend, you—or someone—will probably have to defend again. In inviting you to fight the forces that cause climate change, I'm inviting you to fail repeatedly. In recommending that you work with other people, I recognize that you'll be let down and let others down, in strategy and in compassion. But I would not ask you to join me if only failure were possible.

In the fall of 2019, when construction on the LNG plant had already begun, we learned that another toxic industrial site was pushing for permits on the Southside: a dump for demolition and industrial garbage, with nearly two hundred additional trucks due to lurch on and off the highway every day. Back we went to the city's commitments, the state's regulations; back we went to the spring of anger welling up in our chests and at the roots of our tongues. That January, Monica led a rally outside the city offices and we crowded into a planning board meeting, surging and shifting in the rapidly warming room. The board refused to let us testify, putting their decision off to a later date and increasing our resolve.

I remember that meeting extra clearly because it was one of the last unmasked indoor groups I was part of before it became clear that the coronavirus wasn't going to blow over. The next meeting was scheduled for March 2020. But before that date arrived, the company withdrew their application, and I will always wonder if they got a call from someone on the city planning board, someone who'd witnessed the last fight and saw us pack that meeting and said, *They're going to give you a hard time about this. It's not worth it.*

We didn't stop the LNG plant, but we delayed its opening for two and a half years, and other people—enemies and fellow fighters and future fighters elsewhere—watched us do it. "It brought state agencies to their knees," Gina Rodríguez said. Upstairs, Julian was getting the kids ready for bed. "They are aware of their own regulations that they didn't follow in the past. The city was behind us and I think they are looking at the Port differently knowing that there will be people protesting" the next poison, and the next, "every moment of their day."

Living with the uncertainty of climate change also means living with the uncertain effects of our efforts to slow it. Allowing ourselves to mourn the temporary failures and celebrate the short-term wins are both essential to staying involved and moving. We also benefit from imagining the world beyond the fight. If I'm going to put on the line not just my body but my *life*, the quality and texture of my familiar and beloved life, I want the potential to bring forward a world that contains satisfactions and beauties, puzzles to solve and creatures to pay attention to, love to be upheld by and to carry out.

CHAPTER 8

WHAT HAPPENS NOW

FROM DESPAIR TO (RE)IMAGINATION

YOU ARE HERE

This chapter invites you to learn these things:

- The many livable aspects of our world can teach us how to make it more livable still.
- Even hopelessness can teach us what we need and what we might do.
- Living in uncertainty also means we're living in possibility.
- Our visions of the world we want can help us move—with purpose and with care—beyond the world we know.

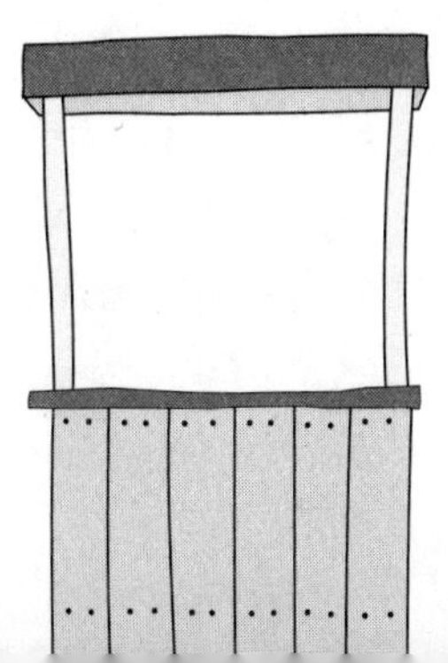

HOW THE WORLD SUSTAINS US NOW

After we learned that National Grid was going ahead with the LNG plant and all the years of damage it would bring, what I felt was despair: a certainty that violence and suffering would prevail, that any effort against them was wasted. The bonds we made fighting the plant had changed our lives and even our city: we fought off the trash company in 2020, funded a generator to refrigerate Monica's grandmother's insulin when Hurricane María hit Puerto Rico, provided groundwork for Providence's Climate Justice Plan and for dozens of smaller acts of community and care. None of that is less real than asthma and cancer and the ever-present threat of disaster. But it's not more real, either. The summer after our defeat, I watched Gina's and Monica's and Andrew's kids running through backyard sprinklers over industrially poisoned grass, cooling off in lead-laced water, and thought how I'd failed them. If I'd fought harder, risked more, left more of myself behind, maybe others would have felt that they could do the same—a different kind of tipping point. I thought how many complicated things would have to happen just for the water to be safe on the kids' skin, the ground safe under their feet.

So few of us have lived in a fully livable world. Even if our own day-to-day has been decent, the systems and activities that maintained it were almost certainly stealing livability from someone else's days and circumstances. And the nasty mix of capitalist economy and individualist culture means that the threat of further unlivability—the lost job, the medical emergency—hangs over our heads even on our good days, even before climate impacts intensify whichever forms of exploitation and domination we were already facing. To imagine living with others in a way where thriving and loss are fairly distributed among all creatures, where there is room for grief and much to celebrate, where rest and plenty are regularly possible for all: this means imagining beyond what we've known, and envisioning ways of caring and being cared for that most of us have never tried.

Yet in other ways, we live in an incredibly, bountifully livable world. Think back to Chapter 3—and if you're able, go outside. What's growing in the sidewalk cracks? What are the clouds up to? The relations of plants and sunlight, air and insects, fungi and soil, animals and water, add up to a world where we can all live, even when some activities also make living harder (diseases spreading, earthquakes forming, humans mining). Most of us have been upheld not only by these relations, but by love and care from other humans, even if some of them have mixed it with hurt. We do have to imagine a world that is truly sustaining, that doesn't tear us down with one hand as it builds us up with the other. But we don't have to imagine what sustenance feels like. We experience it every day, whether we recognize it or not, or we wouldn't be here.

By imagining a world that's livable in *more* ways than the one we know, we give ourselves reasons to make that world. We also envision ourselves as people with the power to make it. Without the fight to stop the sale of Providence water to a for-profit company, the sprinkler water might have been even less safe for the kids to play in. We opened this book with climate anxiety: living with the *what if* of the climate crisis, imagining all the worst *ifs*. Very early in the counseling booth's seasons, a youngish white guy stopped and said that what he felt about climate change was the opposite of anxiety, because anxiety was about something uncertain, and the heating of the Earth was a certainty.

He seemed smug about it. This infuriated me, but he was right, although maybe not in the way that he wanted to be. We don't get anxious about outcomes that we know for sure. We might dread them, or resent them, or be numb to them, but anxiety means uncertainty, and uncertainty—while it's uncomfortable and can be frightening—means possibility.

In a 1994 address to her colleagues and peers, sustainability and systems expert Donella Meadows asked them to imagine the conditions that they wanted everyone to be able to live in. This envisioning gives us both the ability and the reason to work for these conditions. "If we don't know where we want to go," she wrote, "it makes little difference that we make great progress.... The best goal most of us who work toward sustainability offer is the avoidance of catastrophe. We promise survival and not much more. That is a failure of vision." We can follow her urging to describe, fully and in detail, not the world we think we can achieve, or the world we are willing to settle for, but the world we truly want.

Developing Your **If**

Like the other exercises, this is a true practice: it gains strength and flexibility the more you do it. Remember to picture a world that sustains everyone, not just you, and be open to imagining beyond what you have known.

QUESTIONS

If the well-being of the place where you live was complete, what would it be like during the day? At night?

What are some of the things you'd be doing, in that day or night?

What are the relationships between global well-being and this specific place?

PRACTICE

Write or tell the story of your life, starting tomorrow, if one of your "ifs" were true. ("If the water sprinklers were safe for kids . . .") Start by making the vision vivid and thorough: what it looks like, feels like, smells like. Address the *how*, the *who*, and the *where* as they come up: how your medication reaches you in this vision, where the food you eat is grown and who grows it. Describe where the things that used to make you anxious went, and how that was accomplished.

If you're doing this as part of a group, add to and riff on each other's stories. Make your additions reparative, not just critical—for example, if someone's vision of no-car cities would be great for everyone *except* people who use mobility devices, make a suggestion that will allow for wheelchair vans and adaptive trikes. This is also a great one to invite children into! You may need to change the questions a little for younger kids, but their additions to the dream can bring surprising practicality, as well as goofiness and delight.

The balloon of your dreams will always have a string, because you can't think of a world that has no pieces at all of our current one—and some pieces of our current world are nourishing, delightful, and effective as they are. What if the composition and effectiveness of your pills were the same, and they were free, but more people worked fewer hours for more money, using less-polluting processes, in order to make them? People in various places at various times have accomplished parts of that, which suggests that the whole of it is possible. The strength and the weakness of what we imagine is that it's not fully real yet. Bringing it out of what's real now requires understanding the differences between this world and that one, and making generous room for our frustrations, our griefs, and even our despair—but asking them to come along with us, rather than staying stuck with them.

LIVING IN UNCERTAINTY

With every year that passes, climate change adds to the challenges of making the world more livable. It would be dishonest and unkind to suggest otherwise. If the heaviness of the climate nightmare—whether it's in your past, your present, your future, or all three—weighs you down, know that it does the same to me, after all this time and all the ways I've changed. And I feel for you if you find it hard to imagine speaking up at a public meeting, signing away some of your wealth, devoting an evening hour to regulatory research, taking a professional risk by dissenting at work or a physical risk in direct action—especially not knowing whether it will make a difference.

Early in writing this book, I spoke with Faith Kearns, whose book *Getting to the Heart of Science Communication* guides scientists in revising the ways they listen as well as talk about their subjects, especially politicized and high-stakes subjects like climate change. Several years ago at a conference, she watched a fellow scientist display a chart like a thermometer: it showed the range of temperatures in which humans have evolved, lived, and thrived. We haven't, the presenter said, evolved to deal with life outside that window. Our physical bodies weren't "meant" for the temperature window that's coming.

Faith told me about her own research on fire and water systems in California, and her constant awareness that the fire could move at any time from being her topic and her fear to being her reality. She also told me about her nephew, who's in his mid-twenties and lives in rural Arizona: "His fatalism is much larger than mine," she said. "He and his friends have this much more 'who gives a fuck' thing going on about all of it. They can't even picture—there isn't a lot for them to look forward to. I feel that way half the time too."

While carrying that feeling, Faith continues her work of connecting communities with research and resources about fire and water in their landscape. But when you do feel that way, it can feel good—sort of—to hear someone else say it. To put into words the opposite of imagination: the numb, motionless place you turn away from. You can make your life in this place, if that's what you decide. Faced with the dread of large-scale displacement, social breakdown, and bitter weather; the loss of comforts and dreams you've worked for or hoped for; threats to the grip on survival you're barely maintaining—it is tempting to let the climate crisis, and what you believe about the world it's interacting with, make you into a small, mean fortress. But even the fact that it can feel good to hear that temptation put into words is a guide to moving through and beyond it.

Our feelings of despair, says disaster social worker Caroline Contillo, are not to be ignored or dismissed. In a 2022 Twitter thread, she asked people to instead consider that emotions evolved in a relational way to help us cooperate. If that's the case, what can a person's climate despair signal to the people with them? If you don't rush to silence people who bring that feeling forward, what might you be able to learn? "What if we turned to each other and asked, 'What kind of support do you need to come out of this frozen sense of collapsed despair?'" she asked. "What if we become familiar with the people, places and activities that help us move through that frozen collapsed despair without having to jolt ourselves out of it with fear or overstimulation?"

Somatic coach, grief worker, and educator Selin Nurgün steered away from her hopelessness for a long time. In her search for a way to care for the world, she started with fast-paced, constant, driving direct action and a kind of relentless positivity; earned a master's degree in environmental behavior, education, and communication, trying to understand what motivates "pro-environmental" action; took a job encouraging and promoting sustainable purchasing; and became a kayak guide in the waters of Puget Sound. "I noticed people opening up" on those trips, she said, "sharing their stories, their grief, their questions. And I thought to myself, wow. I want more of these conversations." As she began to deepen her own somatic (bodily) awareness and capacity for discomfort, she also began to make room for grief and even hopelessness: to ask what she might learn and receive if she let the hope of retaining certainty, control, and even safety temporarily drop away. "Really hopelessness is like a portal," she said when I reached out to her for an interview. "What is through there? You might think it would destroy you, but what if it doesn't?"

"I love that," I said, though I was also a little shaken. "Like, who will I be, with other people, if this doesn't destroy me?"

Any action that we might take to meet the climate crisis in a real, responsive way requires us to respect despair's presence, as a signal of loneliness and difficulty, but to let go of despair's certainty. It's not just entering the feeling—it's allowing ourselves to imagine, without being sure, what's on the other side.

"What do you do after the emergency?" playwright, poet, and grief guide Diane Exavier asked. We were talking about the end of her mother's life and the end of the world as we know it. (Because Diane is Black in the United States and I am white in the United States, we know it differently.) "You decide to open the door, break down the border, but there's the after of that. How are you going to live?" The border might be the invented border that governments use as an excuse to dominate and kill, or it might be the inner wall you put up to protect yourself from a change that feels like a loss. The door might be the door to your heart or the door to the street—or both—or more. In imagining a different world, you give yourself the gift of imagining a different you, among others, able to meet it. Who do you want to be—the self that is grim, crouching, barricaded? Or the self that takes risks of compassion, sharing sorrow and joy, even within tough conditions? What would you have to let go of, to become that person? Transforming ourselves to meet the transforming world means leaving some aspects of ourselves behind.

"People think the work is being innocent," Diane says, "but actually the work is knowing your part in a well-oiled death machine. Knowing that there might be parts of me that need to die too, the parts of me that keep the machine going through complicity or complacency or ignorance or inaction. And I will welcome those small deaths because I want to keep you and me and us alive."

The greater your investment in white supremacy, in capitalism, in the other hierarchies and structures that feed into climate change—likelier if you're white, possible even if you're not—the more obstacles to changing itself *or* the situation your fearful mind will embrace. Fear is a habit, and a powerful one. So this is also a time to notice how your fear of a future that doesn't work the same way as the present feels in your body—which doesn't know if a threat is real or not. You can fear losing what you have, no matter how much or how little it is, or whether or not it's something that's fair for you to have. You can fear being who you've never been, no matter how you feel about the person you are now. The focus here is not whether the fear is justified, or even whether it's already been affirmed by the world as it is, but how it leads you to act, because we act how we've been practicing. You can practice something else.

Maybe you won't become the person who builds or demands better unless the climate crisis leaves you no choice. That has already happened to many people, and they can be our guides. "I'm a single homeowner and I'm kind of scared," said the woman who visited the Climate Anxiety Counseling booth in the island town of Newport, Rhode Island. "Should I start getting sandbags? Should I get an inflatable raft and keep it in my garage? Am I being paranoid? Am I being silly? Or realistic?" As we talked more, she realized that she had the answer to her own question: "My family back in Anguilla, when those two hurricanes hit, we were so worried, we didn't hear from them for a week and a half. And when they finally got in touch they're like,

'We're fine.' My cousin had food in the fridge and they took the food out and they had a barbecue for everyone! They knew how to do the manual labor, they knew how to put the houses back together.

"The thing that I find different here," she added, meaning the United States, "is that people here are all about that profit—people are gonna be thinking, 'What's in it for me?'" But you don't have to be like that. You can prepare yourself to be changed, so that you can lead yourself and others to be kind—not placating, not condescending, but kind. You can act bravely—not egotistically, not repeating exploitation or control, but bravely. And by doing this, we can make the unfolding present as livable as we can, in ways that are also compatible with a livable future—not a shining city, not a bunker either, but a day-to-day life that we can adjust and enjoy as well as endure. If you have hardened your imagination to protect yourself from what you fear in the climate future—and/or from memories of what you've endured—this exercise will help you make it a little more porous.

Softening the Landscape

Haitian American artist and papermaker Rejin Leys, who taught me this mapmaking practice, started doing it herself and with her neighbors in Queens during the escalation of the "wall" between the United States and Mexico. "Every drawing is so different," she said. "Everything is in flux. If people look at [the maps] and see the US but not the shape that they know, but they recognize it and it's different every time, does it permeate their consciousness? This placement of the border won't always be there."

QUESTIONS

What are your unquestioned assumptions about what climate change is bringing but hasn't brought to you yet?

What are your unquestioned assumptions about what you're capable of doing (or enduring) that haven't been tested yet?

What are your unquestioned assumptions about what's essential to the place where you live that haven't been challenged yet?

PRACTICE

On paper, digitally, or with help, draw the outline of the country, continent, or island where you live now, or where your roots are, from memory and without looking at the drawing (whoever is drawing should look above or away from the page or screen instead of directly at it, and move their drawing tool—pen, pencil, mouse, stylus, or hand—without watching it). Whatever created that outline—river, theft, war, mountain range, treaty, ocean—draw it the way you know or imagine it now.

Now look at what you've drawn. It may differ from the outline you've seen on maps made by other people! Still looking at it, trace your drawing tool around the new "border," not worrying too much if the lines match, cross, or overlap—just follow the general path, going around several times. As the lines accumulate, reflect on the ways your answers to those three questions might shift and change. If it feels appropriate, give thanks and bid farewell to the shape of this place as it once was, and the ways that it did serve or shelter you.

Doing this together, possibly with music, can be companionable. Be kind about each other's maps.

In the summer of 2019, I joined Rejin on Governor's Island, one of the low-lying islands that forms New York City, to invite passersby to draw the borders in their minds and reflect on sea level rise, migration, climate anxiety, and time. Taken aback at first, people gradually spoke more about losses they were mourning and changes they were seeing: a new sandbar island in the Great South Bay where one visitor's uncle had worked in the fisheries for years, a family hut on a cliffside in Greece that was starting to lean after torrential rains. "There's a crack in the cement," that person said, "and this year, everything is falling off."

"I think about the oceans rising a lot," Rejin told me between visitors, "and how the way we think of the world, the geography of the world, won't always be like this. Places that we think of as being there for a long time or permanent—coastlines, borders, cities—they're not. . . . Maybe now is the time to imagine the ways that the world would be different, or better. Maybe making things different or better now, without waiting for the ocean to cover everything."

FINDING YOUR OWN EDGE OF POSSIBILITY

When you do the first exercise in this chapter, pretty quickly you find yourself imagining structural changes, even if your starting point was close to home. Getting that medication turns into not just Medicare for All, but worker-owned labs and an end to price gouging; clean water for the sprinklers involves not just massive investment in the water utility itself, replacing pipes and other equipment, but returning the reservoirs to local Indigenous care and dismantling industries that keep the water full of poison.

Sherrie Anne Andre, whom you met in Chapter 6, learned to climb right after a blizzard. Climbing with ropes and pulleys meant they could hang a TOXINS ARE TRESPASSING ON OUR BODIES banner from a bridge or building, reach a platform to delay demolition workers from felling a tree, or mount a tripod to obstruct a prison driveway. "My knees were swollen from pressing up against this frozen tree for hours," they recalled of their training days, "and I had this emotional response to harming my own body." And that was before they joined water protectors as a medic at Standing Rock, blocking the Dakota Access Pipeline in freezing weather, faced with armed police and National Guard troops; before the State of Massachusetts arrested, tried, and jailed them for thirty days for blocking the entrance to an ICE facility. "A lot of times activists are mentally putting aside our trauma to prepare to be harmed by the state. And that really fucks with you."

But at the same time, they said, there are ways it feels right. "I think for some people it is a way of getting across our feelings about what we're experiencing in our communities without having to talk about it. Especially for BIPOC people who feel that their ancestors hold a warrior spirit or warrior intention, it is a way for us to channel our energy into these types of tactics. We don't have to tame ourselves so that we are being heard."

Direct action of this kind often interrupts "business as usual" and includes criminalized actions like obstructing highways, locking down to construction equipment so it can't be used, or destroying property. You might remember the Valve Turners, several groups of activists who at different times and sites have accessed and turned off emergency shutoff valves to stop the flow of oil, or the people who destroyed construction equipment in defense of the Weelaunee Forest. When Sherrie joins others in physically blocking a pipeline route, they're literally slowing down ecological damage. They're also illuminating how careless of life the fuel companies and investors are: when prevented even for a moment from the slow violence of pollution, those who profit from it escalate to active violence.

Sherrie pointed out the strategic use of bringing that escalation to light: "As a person who doesn't have a lot of social capital, my body is all I have. I can't hire a lawyer to file lawsuits, I'm just not famous enough to be heard." They laughed. "I'm not Yoko Ono, no one's listening to me about stopping fracking. This is all I have, and it's something that other people can relate to because they have it too! Like, 'Oh, somebody put their body there'"—in immediate harm's way, in the path of greater damage. "'Why would they do that with their body? I wouldn't do that with my body!' So even if they don't agree with me, they will have some sort of feeling about what I'm doing."

Does this resonate with you? Can you find this nature in yourself? If you do have the money to file lawsuits against fossil fuel companies, or if you are famous enough to be heard, placing those powers in the service of climate justice is a great move. Longtime organizers often point out the usefulness of *diversity of tactics*, meaning not only that there are many ways to work toward our shared survival, but that using several at once can create a better chance of success. If (like the majority of us) you have neither fame nor surplus money, the stories and exercises in earlier chapters and in this chapter's next section offer plenty of other options for connection and action, but I ask you to imagine this one, too. The point is not that everyone can or needs to do what Sherrie has done, but that you, like Sherrie, can find a place to meet the climate crisis where what's effective, what's possible, what feels right, and what *is* right converge.

What's not right is for the same people to take on the hardest roles, over and over and over. The nonviolent direct actions, and in some cases the violent actions, too, of people whose "bodies are all they have" in defense of their lives and the living world—often Black people, Indigenous people, people of color, and disabled people—are brave and powerful, and have been effective. That courage is worth honor and celebration, alongside mourning and rage that it was necessary at all—there should have *been* no threat, no harmful occasion to rise to. And those who do rise to it, by choice or by necessity, could use some more company. More people taking part in direct action of this kind makes it more effective. It also means that more people know what it's like—what it takes and what it offers. Their responses to it will be rooted in experience. If they are afraid of it, they'll know what it is they're afraid of. Their recognition of its power—collective, shared power, backed up by planning, supported by love—will come from feeling it. And they'll know that it's possible to directly limit the forces that are hurting us.

This next exercise asks you to imagine taking on that role—physically interrupting an environmentally harmful process—if you have never done so. Take care to surround your vision with grounding practices, and release your reactions with sound or motion. It's okay to not know the answers to the questions, and to be shaken by the practice; it's also important to feel that the questions are answerable, that this kind of action is thinkable, even if it's not doable by you right now and may never be what you do. If you've taken on this kind of risk and damage before, and don't want to relive it, skip this one.

Being the Barrier

One more thing this exercise is meant to emphasize: it's never 100 percent true that anyone's body is all they have. Your relationships with the living and the dead, your training and skills—you couldn't, wouldn't, and shouldn't place yourself between violence and its target without these. If the warrior role is not right for you, the second and third questions can offer you another path.

QUESTIONS

Where—especially, but not only, near the place where you live—would you place yourself to block some cause of climate or ecological damage? (The driveway of a methane-fueled power plant, the entrance to an oil port, the road for a clear-cutting site...)

What arrangements would you make beforehand with people in your life or group? (Childcare, pet care, legal and jail support, renewing prescriptions, clearing debt...)

What kind of care and support would you want after? (Meals delivered, company for chores, a list of sliding-scale therapists prepared...)

How would you want to be together with the people in your life the night before?

PRACTICE

Design the ritual sendoff you'd want if you were going to take this kind of risk to block or illuminate climate damage. Imagine who'd be present, what you'd want them to say or do; imagine what you'd eat and wear. Name the people whose support or blessing you'd call on, human and more-than-human, dead and living: ancestors, plants, animals, elements, heroes.

If you're doing this with a group, design the ritual together. Rehearse it.

If you're not blocking the road, maybe you can walk the dog, do the grocery shopping, make or coordinate phone calls to the jail, bring clothes to the laundromat, show up to the court date, or donate your other skills and time from counseling to car repair. Again, these roles also require a transformation of your life: who you're in relationships with, how you use your energy and time, and how taking on such a role will serve you, too.

Taking a client or customer for free might mean you lose the capacity for one paid appointment, or the time for one muffler replacement, and you're the best judge of whether you can afford that—not just in money. The time and attention for that interaction might need to be budgeted away from your friendships, your family, your job, or other commitments, which means the potential for more conversation and conflict. But these interactions also enact a world where quality of life matters, where people take care of one another and ease each other's burdens where we can, where responsibilities ebb and flow so that there's some slack and some backup. If, in those interactions, you're making in the present the world you want to continue, then part of what you get in return may be a sense of honesty, of rightness, of satisfaction. The gap between the world that is and the world that could be, the person you are and the person you could become, is closing.

"Direct action is healing work," Sherrie said. "For me when we do direct actions, it's never like all of a sudden I think [a fossil fuel company] is gonna create something for my community that we need or want, it's like I need them to stop! So I can create the resources we need! I think that's one of the reasons we do direct action, is to say, 'We need you out of our way so that we can create something else.'"

CARING FOR EACH OTHER, OURSELVES, AND THE WORLD

You'll remember how, in Chapter 6, Sherrie was also an advocate for building community care into the workings of their organization. As a term, community or collective care was solidified by members of the disability justice movement. As a practice, it's as old as some of the oldest humans we know about today: archaeologists have found the skeletons of ancient people who lived many years after injuries or illnesses that would have incapacitated them, tended in life, buried with reverence in death. And it's as contemporary as the last time you drove a friend to an appointment, listened without judgment, looked up lawyers, dropped off soup—or the last time someone did those things for you as a neighbor or friend. It describes methods and structures of care for which you don't need to pay or apply or qualify, it includes developing and sharing emotional and relational skills like the ones in this book, and it is an essential part of building a livable present and future within climate change.

Moments of practical care like the ones previously mentioned, especially if they're steady and reliable, reduce people's overall strain and isolation: they have less to pay attention to and do, because they matter to someone else. Responses to disaster, from hosting displaced friends and family to community cleanups and water distributions, also overlap with collective care. So do rent parties and eviction defense for people whose homes are threatened, not by climate change itself, but by the exploitative forces that cause it. Joining together in protection for a river, a forest, or a neighborhood is care *for* the community as a living system, as well as for the people involved: think of the bonds we built fighting the LNG plant, bonds that strengthened us even though we lost, and the sense of purpose that our anger and love gave us.

As Deb Krol in Chapter 3, Victor Ibarra in Chapter 5, and others have pointed out, maintaining good relationships with land and water, and making it easier for others to do so, is also care: enabling Land Back and Indigenous land and water stewardship, enriching soil to help it grow food and store greenhouse gases, removing a dam so that fish can freely travel and multiply. Caring for the land is also something that adults and kids can do together, sharing a sense of responsibility as well as knowledge and enjoyment, and learning more together as you go. And for some, caring for the more-than-human is easier and suits us better: you don't need to set a boundary with a meadow the way that you often need to with a person.

Good care requires good boundaries, because you also are a person and it's no good if the only way for someone else to be okay is for you to suffer—especially if that's how it shakes out every time. (That's also a very clear summary of what's wrong with our current systems, and how maintaining boundaries is part of making a more livable world: a situation where you're able to say "no" and have it stick is better than one where you don't.) Being clear about what you can do or what feels wrong is also a gift in the long run to the person or group that needs care or other input from you: it means they won't be expecting what you can't give, and can ask someone else or adjust the plan, rather than leaving undone things that need doing.

Getting better at saying what you can and can't do is also good practice for *receiving* care, and asking for it when you need it. Here's an exercise to help you practice that.

Receiving Abundance

QUESTIONS

What are some forms of care that you easily recognize and accept?

What are some forms of care that you distrust and reject?

When you've allowed yourself to be cared for in the past, and the care reached you, what made that possible?

PRACTICE

Write, type, or record your answers to the first question. Tell them to at least three people, whether or not you think those people can offer them (if some of them are people who already care for you in these ways, thanking them out loud is good to do). You're not necessarily asking for these forms of care in the moment (though you can), just letting people know what you sometimes like or need.

If you're doing this as part of a group, make a note of each other's forms of care, in a way you'll be able to find again and remember. Practice saying yes *and* no to one another.

If you're watching Person A's kids so that he can drive Person B to an appointment, and especially if there's a Person C you can call when you fully intended to watch Person A's kids but need a rest yourself, you've got the makings of what Leah Lakshmi Piepzna-Samarasinha, author of disability justice writings *Care Work* and *The Future Is Disabled*, calls a care web. Not always reciprocal (though they can be), not just transactional (though they can be), care webs interlace through the relationships and systems of our lives to try to hold us up. Many people already need and have and are part of them; more of us are likely to need them more as climate change and the forces that cause it make the larger systems we rely on more brittle and likelier to fail us.

Sometimes care webs also fail, of course—we fail. Sometimes the breakdown is in one or more threads, sometimes in the relation between them, and the costs can be high and add up. Redundancy (more than one person who can do the same things) is useful; if your web involves more people who can do a little, it'll be less common for anyone to push past their own limits and easier to build acts of care into the rest of your lives and obligations.

In their books, Leah Lakshmi Piepzna-Samarasinha quotes disabled friends who don't ever want to have to depend on friendship or goodwill to take a shit and get cleaned up. Reliably maintaining that simple dignity, without exploiting anyone else, requires a Just Transition so radical that it's beyond anything we've known, and so pushing for better systems on a larger scale becomes a form of care as well. Sometimes our care webs fail because of familiar enemies, structures of exploitation and domination—you can turn a vacant lot into an activity park for neighborhood kids, or a safe camping place if you can't afford housing indoors, or a community garden, and someone can decide you're a threat and demand that the state chase you out. That's where Sherrie's point about direct action as care comes in: blocking those threatening forces so that care can go forward and healing can begin. I think it's also where Selin's thoughts about hopelessness are tending: if we stop hoping that the things we know don't work will *start* to work, or stop hoping that we'll be able to hold on to lives or dreams we can't keep without hurting other lives, we may open ourselves to true transformation—of ourselves, our systems, and our world.

In order to be brave, in order to stay kind, within uncertainty, precarity, and risk, we are going to have to offer and accept more kinds of care. You can get better at this, too. Practicing and sharing the methods this book offers—for steadying ourselves and each other, tending grief, directing anger, finding rest, and riding the wave of emotion into connection and action—are just the beginning of the skills, methods, and structures out there. There are more within you already, and still more in the resource section at the back of this book. These methods are also the beginning of a world where more people have the emotional and relational expertise to forgive themselves, and others, when their efforts to build a more just and nourishing world fall short. A world where people can take the risk of being present with despair and grief and rage, their own and that of others. A world where people who've been hurt have ways, and reasons, to punch up instead of down. A world where one of those people is you, and some of the others are people you know.

The uneven distribution and pace of climate impacts, and the variation among people who are using this book, mean that I don't know which wounds you're going to tend, what joys you're going to find, or which path you're going to widen by walking or rolling along, or by providing shelter at its side. But I am certain that the next part of your path is waiting for you.

BECOMING WHO YOU WANT TO BE TOMORROW, TODAY

When people ask, at the counseling booth and elsewhere, "Is the world about to end?" they often mean, "Does the climate crisis mean that my life—my life as I've known it, or the way I want to live it—has to end?" Between things that will become less possible to do, things that it will be less ethical to do, and things that differ from what you're doing now but will actually *improve* your and others' lives as the climate changes, the answer may be yes.

But after your life as you know it ends, as long as you're still physically alive and emotionally present, there's the uncertainty (which, remember, means the possibility) brought by another day, and another. The one certainty that comes with us when we're born—that we will die—is neutral, just like being born is. It's how, it's when, for whom, for what. Who do you want to be, with others, when the changes come that end your life as you know it? Who will be with you when you move out to meet them? What vision of the world will you be contributing to, just by making that move?

In both of their recent books and from within their lifework of disability justice, Leah Lakshmi Piepzna-Samarasinha describes the tiny, daily acts of alertness and care, laughter and grace, that make and enliven their life and the lives of their sick and disabled kin. They also propose "cripping the Green New Deal," ensuring that a Just Transition of work, at the necessary scale, includes people with all degrees and kinds of illness and ability. And they further imagine making beautiful and challenging art, sharing adaptive technologies through free libraries, and the "glorious lipstick-red spiral ramp" they want curving around the home they don't yet have, for them and their friends who use canes, walkers, and chairs to find ease and celebration in the ways they must move. They're building on realities—the library and the art are both real, just not common—to diversify *imaginative* tactics: assessing both dream and actual choices and activities for their ability to contribute to more good days for more people.

Remember Tiffiney Davis and her dream of a community hub? While the Red Hook Art Project doesn't have a building of its own yet, its new premises have a free store and meal distribution as well as art classes—and it's sharing meals with the refugee camps that the city has placed in the neighborhood. Everyone who visits, and as a result has more to share with the people around them, will be making the world. Cooperatively owned solar micro-grids are starting to supply electricity in parts of Puerto Rico, cutting out the need to pay the price-gouging and unreliable utility company: maybe next time there's a hurricane, Monica's grandmother won't need to run a gas generator to keep her insulin cold and usable. Everyone who participates in those solar co-ops is making the world. Dezaraye's students, learning to steward the land in the Central Valley and their own labor, are making the world.

When I thought about the fight we lost and all the other injuries to life on Earth and the ways I'd personally failed, I wasn't wrong. What I was seeing was there. But I wasn't allowing myself to feel everything that was there, all the remaking of the world that was going on around me. And I was forgetting that others—people I loved, people I didn't even know, human and more-than-human beings—were taking up the work of that remaking, that my place in it was to be useful but small, and that I could dance (or wobble) into and out of and back into that place.

Sometimes we find and join fully formed collective efforts. Other times we nourish little pieces—a peer support network here, a restored wetland there—and look for the ways what we're doing complements what others are doing elsewhere. Sometimes the changes we work for don't seem related to climate change at all. And yet because we're interdependent with one another, relieving strain or building strength in one part of a system changes the weight and the dynamics in other parts, opening up availability and capability for others.

For our lives to satisfy us—however long and however safe they are—we need to both feel like our actions have a purpose, and be right about that. So far, I've avoided using the word *values* in this book—I don't like its implications of property and worth that seem to put a price tag on what matters. Maybe it's *mattering* that I can ask you to notice: the moments, within this profoundly disrupted and disturbing time, where it feels like your attention and effort are where they belong, on the things that matter to you and not only to you. That is the feeling of rightness within wrongness that Sherrie was describing in their warrior actions; that is the feeling of rightness I hold when I tend plants and soil at a community farm, or watch a red milkweed beetle with the kids, or lead a group of high school students or union members in an exercise that lets them hear each other about the vulnerable beloved places where they live. And there may be fights I join and risks I take that are just over the horizon but that I can't see yet, where I have a chance to make a different set of choices and embrace the possibility of different losses. I look beyond my desire for certainty—that such-and-such climate action will "work," or alternatively, that "we're all doomed" and so what I do doesn't matter. I fear that, but I don't believe it. I look up, and I see you.

I can't promise you a life where no one will hurt you, but you don't have that now. I can't promise you a life where the rules won't change on us, but we don't have that now. I wish I could say for sure that you'll feel better *and* live better if you do what I recommend, but that too is uncertain. I do know that we weather hardship, disaster, and even violence better, and find more reasons to summon courage and endurance, when we know other people and other people know us. We also have people to celebrate with: the big victories, the small delights, the openings to keep going. Love, conflict, care, and joy are all things it's possible to get better at through practice, and we *can* change to become part of what our communities need. I did. I will again. Between those changes, right now, where all of us live, I offer you one more set of questions and one more practice to help you inhabit the climate present, move into the climate future, and become yourself.

Acting on Love

QUESTIONS

When you are your bravest, kindest self, what do you want your response to climate change to be?

Whom do you want to invite to join you?

Who can help you grieve without sinking into despair?

Who can enjoy and cherish the living world with you?

Who can enable your deepest and most thrilling visions of possibility?

PRACTICE

Make a date with at least one of those people, and keep it.

HOW THIS BOOK CAME TOGETHER

NOT LONG AFTER I STARTED sitting out with the Climate Anxiety Counseling booth, people I knew started asking me if I was going to write a book about it. Through several seasons, and through expanding my responses to climate change to the level of community participation and action, I'd learned a lot, including many things that seemed worth sharing. But even with others' perspectives added, mine felt too limited, and I didn't know what I'd be asking readers to do with what they'd read—because I didn't really know what to do with it myself. When I realized I was repeatedly asking different people some of the same questions and making some of the same recommendations, I detected another piece of the possibility, and I started forming those conversations into some of the exercises you've just tried. But I'd *done* so little, comparatively. I wanted you to hear from a range of people who, taken all together, had done more than any one person could do, and could offer you a variety of models for what to do next. And I wanted those models for myself, too—what could *I* be doing next?

Some of the people I reached out to for longer interviews, I already knew: they'd talked with me at the booth, organized me for local climate and environmental justice action, or presented at a series of events I convened called Interdependence Days (long story). Because I also talked about the booth on Twitter and followed people who I thought might help me do it better, I was at least slightly connected to several people who had powerful things to say about what they were surviving, studying, or enacting, or could connect me to others with things to say. That was trickier, because there is of course a long and bad tradition of writers coming into places where difficulty of life has been structurally induced, wringing people out like dishrags for their stories, and then leaving, with no benefit to them.

I thought of two ways to minimize that: If I was reaching out to strangers, I chose people who'd already spoken or written publicly about their experiences, showing that doing so didn't feel off-limits for them. And I included in my request the offer of something in exchange, either money or a trade of services, like help with chores (for someone local) or, say, copy editing (for someone far away). An early interviewee pointed out that being very specific and clear about this second thing was important, even to the dollar amount: people deserve the option of informed consent, and to do that they need to know whether it's a "real compensation" situation, a "nice gesture" situation, or a "donating my time and knowledge" situation. People also deserve to have the final say over how their words, ideas, and story appear, which is something that I did build in from the very beginning: the earliest email exchanges also include a commitment that interlocutors would be able to review how I quote them and what I write about them, request changes, and have those requests honored. I also stuck to this for booth interlocutors whom I could reach.

The book is very US-centric, which is a limitation—my limitation. That's where I've always lived, where the ecosystems are that I'm within, where I have such cultural and systemic competence as I possess, and where my relationships are. Almost everyone I spoke with either lives on this continent or did at one point. When I name places, I've tried to indicate both how settlers have renamed them and the name of the people indigenous to those places, at least in the initial mention. The people whose lives on these lands were violated and interrupted are often there, still; the people who benefit, still, from that violence and interruption are also there. This is something to talk about, but not *just* to talk about, as Chapter 3 makes clear. The same is true of the ways that climate change's two main causes—capitalism and white supremacy—outsource the climate damage they do to places outside the United States' made-up borders, places that they've already dominated and impoverished. I have included in the resources (starting on page 207) some writing not just about how this happens but about how people in those places have fought back against it, and sustained one another in those fights. And it still seems to me that no matter where you end up, you have to start where you are, talking with the people you're with.

Not every longer conversation I held made it into the book. The people you're hearing from and the groups you're hearing about are here not because they offer a perfect roadmap for living within climate change or enacting an alternative to the systems that cause its (and so many other) harms, but because—in relationship with many others, unnamed here—they demonstrate facets, pieces, knots in a net that can hold us up. The whole point of this book is that no one person can do everything that needs doing in this moment. Elements of what these interlocutors and their organizations are doing are good now and could be good later, or could be mixed and matched to create something we can all, literally, live with.

I tried to identify and place people in the ways that were relevant to or would show their authority in what they were saying—work they have done, relationships they're in, places they've lived, as well as elements of their life like their race, ability, or economic class that have affected all those elements: how they live, how others treat them, and what stands out to them. I also sometimes stated someone's race if it was part of what they told me about themselves, even if that doesn't show up in the quote or story, because it was important enough to them to mention, and in those cases I went with the terms they applied to themselves. I was inconsistent about this, both because not every story discussed here is enriched by that kind of context and because not everyone's identity is known to me, especially people who spoke to me at the counseling booth. And people have other aspects of their identity that neither they nor I mentioned that may inform their perspective—this also applies to me. What we have lived affects what we know and feel; by bringing together these particular experiences, both individually and structurally informed, we gain not just authority but knowledge.

When Hachette (shoutout to Renee Sedliar) bought the book and sent me and my agent (shoutout to Jen Marshall) the first advance check, I realized—I'm sorry this sounds so sanctimonious but it really is what happened—that sharing it with the people I'd talked with, especially in the long interviews that I'd sought out, was the right thing to do. The book simply would not exist without the conversations we held and the time and work they generously offered me. And because of some things that are true about me (I have a high-paying job, my medical expenses are low, I have no financial dependents, and due to generational wealth I have no major debt) I didn't need to keep all of my share of the money.

I did some math, and settled with contributors on the amounts that felt, to them, substantial and respectful. But the money was only part of it; the other part was making sure that their words, ideas, and stories were going out into the world in a form they approved of. Later in the process, when the draft was closer to its final form, I began sharing with all named contributors the chapters where they appeared, with their sections highlighted. I offered the options of indicating their satisfaction with the chapter as it was, requesting changes, or withdrawing their participation. I also extended my offer of a share of the advance. Some people were happy with their contributions as is, or with quick changes, like correcting the spelling of a child's name. Others went through a longer process with me to make sure that they, their relationships, and their activities were represented precisely.

Reaching out to people was exciting and also hard. In some cases, I'd dropped the ball on what had been an ongoing communication, and was relieved when they responded with grace. Still others had concerns about consent and ownership of their material that took time and care from both of us to resolve. I'm glad we went through this process. And I was so happy to hear what they did with the money. Some places that parts of the advance have gone: into the commissary account of an imprisoned land defender; to support an Indigenous land trust; to meet the contributor's own needs; as a monthly donation to an organization that supports people dealing with state and border violence; to get a kid with disabilities to a camp he loves; and to help a young Indigenous woman hang on to family land. If the book does nothing else good for anyone, ever, it did this.

Obviously, my intention and desire is that way more people, including you, will do way more good things with it. This book is not the last word on anything—or the first word either. When Blia Moua (from Chapter 2) signed off on his contribution, he pointed out that as I shared this book with others I would learn more about what else was needed and improve upon what was in it, and I am taking that to heart. The book is a continuation, a development, an attempt, and I hope to stay in contact and relationship with everyone who's interested in continuing to try. If anything in it proves to be useless, it can be discarded; but also, I look forward to the moment when some of the elements of this book that are useful or resonant now can also be built upon, pierced through, or spun into something even better. When I and you are even better at becoming caring and purposeful parts of the relations we're in and the world we're making. When there are more and better stories to tell, and to be told. Thank you for coming with me this far.

LEARN MORE

CONTRIBUTORS AND RESOURCES

IN THIS LIST, YOU'LL FIND:

The (named) people who spoke into each chapter, either at length with me or in earlier work that I quoted, in order of appearance.

Titles (and links) to work that informed this book, or will guide you beyond this book, in alphabetical order by family name or name of organization.

It is not a comprehensive list, any more than the book is a comprehensive book, but it offers you some ways to expand and complicate your understanding of the practices and possibilities that mean the most to you. (It doesn't include much audio and video media because I observe those only if I must.)

Some of these are online; others you can buy if you have the means, request from a nearby library, or ask if a friend wants to share. Like this book, each holds pieces, elements, components that you can read with a careful mind, put into relationship with one another, and apply as you meet the world.

BODY-MIND PRACTICES: FROM SOMATIC EXPERIENCING TO ORGANIZING

CHAPTER 1

Nicole Hernandez Hammer

Caroline Contillo

Selin Nurgün

Britt Wray

Nicole Capobianco

Tiffiney Davis

Augusto Boal, *Theater of the Oppressed*
Aria Boutet's Instagram: https://instagram.com/queencrones
Staci K. Haines, *The Politics of Trauma*
Resmaa Menakem, *My Grandmother's Hands: Racialized Trauma and the Path to Mending Our Hearts and Bodies*
Yoko Ono, *Grapefruit*
Red Hook Art Project: www.redhookartproject.org/
Todd Shalom, *Prompts for Participatory Walks*
Jeeyon Shim's games: https://jeeyonshim.games/
Britt Wray, *Generation Dread: Finding Purpose in an Age of Climate Crisis*

MUTUAL AID AND COMMUNITY ABUNDANCE

CHAPTER 2

Eva Amanda Agudelo

Out of the Woods Collective

Blia Moua

Angela Blanchard

Cindy Quezada

Nwamaka Agbo, Gopal Dayaneni, Karissa Lewis, and Robert Hawkstorm Bergin: https://centerforneweconomics.org/publications/a-conversation-about-land-and-liberation/

Mia Birdsong, *How We Show Up*

Cooperation Jackson: https://cooperationjackson.org/

Hope's Harvest: www.farmfreshri.org/programs/hopes-harvest/

Natalie Ironside's Tumblr: www.tumblr.com/natalieironside

Errico Malatesta, *Anarchist Essays and Thoughts on Mutual Aid, Crime, Government, and Organization*

Premee Mohamed, *The Annual Migration of Clouds*

Multisolving Institute: www.multisolving.org/

The Othering and Belonging Institute: https://belonging.berkeley.edu/

Out of the Woods Collective, *Hope Against Hope: Writings on Ecological Crisis* and "On Climate / Borders / Survival / Care / Struggle": www.basepublication.org/?p=474

Leah Penniman, *Farming While Black: Soul Fire Farm's Practical Guide to Liberation on the Land*

Sustainable Economies Law Center: www.theselc.org/

HUMANS WITHIN LIVING SYSTEMS

CHAPTER 3

Don Hankins

Diane Schapira

Diane Exavier

Deb Krol

John Borrows

John Borrows, https://raventrust.com/john-borrows-on-mindfulness-and-indigenous-law/

Frantz Fanon, *The Wretched of the Earth*

Alexis Pauline Gumbs, *Undrowned: Black Feminist Lessons from Marine Mammals*

Jessica Hernandez, *Fresh Banana Leaves: Healing Indigenous Landscapes Through Indigenous Science*

Intergovernmental Panel on Climate Change Report (2023): www.ipcc.ch/report/ar6/syr/

Jamaica Kincaid, *My Garden (Book)*

Patty Krawec, *Becoming Kin: An Indigenous Call to Unforgetting the Past and Imagining Our Future*

Land Back: A Yellowhead Institute Red Paper: https://redpaper.yellowheadinstitute.org/

Lynda V. Mapes and Philip-Daniel Ducasse, "Course Correction": https://atmos.earth/yurok-tribe-restorative-justice-klamath-river

Zoe S. Todd's body of work: https://fishphilosophy.org/

Amber Webb, Dr. Allison Kelliher, Holly Nordlum, and Melissa Shaginoff, "Decolonizing Bodies": www.youtube.com/watch?v=_9NB9QSJ034

Andreas Weber, *Matter and Desire*

ECOLOGICAL GRIEF: CELEBRATION AND LETTING GO

CHAPTER 4

Dezaraye Bagalayos

Joanna Macy / Work That Reconnects

Maya Weeks

adrienne maree brown

Britt Wray

Rob Hansen

Allensworth Progressive Association: www.linkedin.com/company/allensworth-progressive-association?trk=public_jobs_jserp-result_job-search-card-subtitle

adrienne maree brown, *Emergent Strategy: Shaping Change, Changing Worlds*

Octavia Butler, *Parable of the Sower*

ShaLeigh Comerford, "Memos for Migration": https://ecotonemagazine.org/various-instructions/memos-for-migration/

Martha Crawford, What a Shrink Thinks: https://whatashrinkthinks.substack.com/

Mary Annaïse Heglar, "Climate Change Isn't the First Existential Threat": https://zora.medium.com/sorry-yall-but-climate-change-ain-t-the-first-existential-threat-b3c999267aa0

J. B. MacKinnon, "The Whale Dying on the Mountain": https://hakaimagazine.com/features/whale-dying-mountain/

Kari Norgaard and Ron Reed, "Emotional Impacts of Environmental Decline": https://pages.uoregon.edu/norgaard/pdf/Emotions-of-Environmental-Decline-Norgaard-Reed-2017.pdf

WORK: JUST TRANSITION AND DEGROWTH

CHAPTER 5

Nate Lake

Angela Blanchard

Victor Ibarra

Renée Peperone

Rebecca Leber

Herman Daly

Jamie Tyberg

Kate Beaton, *Ducks: Two Years in the Oil Sands*
Eula Biss, *Having and Being Had*
BK ROT: www.bkrot.org/
Climate Justice Alliance: https://climatejusticealliance.org/just-transition/
Sufyan Droubi, Arthur Galamba, Fernando Lannes Fernandes, Amanda Andre de Mendonca, and Rafael J. Heffron, "Transforming Education for the Just Transition": www.sciencedirect.com/science/article/pii/S2214629623001500
Vanessa Jimenez Gabb, *Images for Radical Politics*
Robin D. G. Kelley, *Hammer and Hoe*
Alicia Kennedy's newsletter: www.aliciakennedy.news/
Timothee Parrique's blog: https://timotheeparrique.com
Thea Riofrancos, *Resource Radicals: From Petro-Nationalism to Post-Extractivism in Ecuador*
Jamie Tyberg and Erica Jung, "Degrowth and Revolutionary Organizing": https://rosalux.nyc/degrowth-and-revolutionary-organizing/

FINDING AND FILLING YOUR PLACE

CHAPTER 6

Mark Baumer

Tamara Toles O'Laughlin / Climate Critical

Sherrie Anne Andre

Laura Brown-Lavoie

Kyle Powys Whyte

Mark Baumer: https://notgoingtomakeit.com/
Climate Burnout Report: www.climatecritical.earth/report
Jo Freeman, "The Tyranny of Structurelessness": www.jofreeman.com/joreen/tyranny.htm
Kelly Hayes and Mariame Kaba, *Let This Radicalize You*
Victoria Law, China Martens, and others, *Don't Leave Your Friends Behind: Concrete Ways to Support Families in Social Justice Movements and Communities*
Renee Lertzman, Project Inside Out: https://projectinsideout.net
Tema Okun and others, "White Supremacy Culture": www.whitesupremacyculture.info/
Rithika Ramamurthy and Olúfẹ́mi O. Táíwò, "Constructing Solidarity": https://nonprofitquarterly.org/constructing-solidarity-an-interview-with-olufemi-o-taiwo/
Aimee Lewis Reau, Chelsea Rivera, and LaUra Schmidt, *How to Live in a Chaotic Climate*
Eve Tuck and K. Wayne Yang, "Decolonization Is Not a Metaphor": https://clas.osu.edu/sites/clas.osu.edu/files/Tuck%20and%20Yang%202012%20Decolonization%20is%20not%20a%20metaphor.pdf
Kyle Powys Whyte: www.nationalobserver.com/2019/02/15/news/urgency-climate-change-advocacy-backfiring-says-citizen-potawatomi-nation-scientist

CONFLICT AND ANGER

CHAPTER 7

Gina Mariela Rodríguez

Monica Huertas

Diane Exavier

Weyam Ghadbian and Jovida Ross

Yotam Marom

James Kuo

Weyam Ghadbian and Jovida Ross: https://communityresourcehub.org/resources/turning-towards-each-other-a-conflict-workbook/

Interrupting Criminalization and Dragonfly Partners, *In It Together*: www.interruptingcriminalization.com/in-it-together

Amanda Machin, "Democracy, Agony and Rupture: A Critique of Climate Citizens' Assemblies": https://link.springer.com/article/10.1007/s11615-023-00455-5

Yotam Marom, https://medium.com/@YotamMarom/moving-toward-conflict-for-the-sake-of-good-strategy-9ad0aa28b529

Rabbi Danya Ruttenberg, *On Repentance and Repair: Making Amends in an Unapologetic World*

Samantha K. Stanley, Teaghan L. Hogg, Zoe Leviston, and Iain Walker, "From Anger to Action: Differential Impacts of Eco-Anxiety, Eco-Depression, and Eco-Anger on Climate Action and Well-Being": www.sciencedirect.com/science/article/pii/S2667278221000018?via%3Dihub

The Wildfire Project: https://wildfireproject.org/

DIRECT ACTION: RADICAL IMAGINATION AND COMMUNITY CARE

CHAPTER 8

Donella Meadows

Faith Kearns

Caroline Contillo

Selin Nurgün

Rejin Leys

Leah Lakshmi Piepzna-Samarasinha

Beatrice Adler-Bolton and Jules Gill-Peterson, "Imagine What We'll Build for One Another": https://thenewinquiry.com/imagine-what-well-build-for-one-another-an-interview-with-jules-gill-peterson/

Bayo Akomolafe's body of work: www.bayoakomolafe.net/

Anonymous, "Treasure Hunt for Coastal Gaslink": https://mtlcounterinfo.org/treasure-hunt-for-coastal-gaslink/

Defend the Atlanta Forest: https://defendtheatlantaforest.org/

Peter Gelderloos, *The Solutions Are Already Here: Tactics for Ecological Revolution from Below*

Standing Bear John Gonzalez, *Standing Rock Is Everywhere*

Tara Houska, "The Umbilicus": https://atmos.earth/the-umbilicus-line-three-indigeneity-tara-houska/

Faith Kearns, *Getting to the Heart of Science Communication*

Renee Linklater, *Decolonizing Trauma Work*

Keguro Macharia, "Not This. More That!": https://thenewinquiry.com/blog/not-this-more-that/

Donella Meadows: https://donellameadows.org/archives/envisioning-a-sustainable-world/ (presentation at the Third Biennial Meeting of the International Society for Ecological Economics, San Jose, Costa Rica)

Fred Moten and Stefano Harney, *The Undercommons: Fugitive Planning and Black Study*

Vicky Osterweil, *In Defense of Looting*
Leah Lakshmi Piepzna-Samarasinha, *Care Work* and *The Future Is Disabled*
Project LETS: https://projectlets.org/
Leanne Betasamosake Simpson, *A Short History of the Blockade: Giant Beavers, Diplomacy, and Regeneration in Nishnaabewin*
Leanne Betasamosake Simpson and Robyn Maynard, *Rehearsals for Living*
Jeff VanderMeer, *Annihilation*

ACKNOWLEDGMENTS

EVERYONE WHOSE NAME APPEARS EARLIER in these pages shared with me their insight, their expertise, and their time, with generosity so incredible it makes me catch my breath. I hope you like what we have made and can use it for your own things. I hope it reminds you that you are not alone. Several people whose conversations didn't make it into the final version still shaped and enriched it; thank you to Charlotte Abotsi, Amy Balkin, Megan Brown, Kristen Dobbin, Meghan Kallman, Ramy Kim, Tati Luboviski-Acosta, Elisabeth Nicula, and Molly Whitely for talking with me.

Everyone who spoke to me at the Climate Anxiety Counseling booth, from friends to strangers, shared with me—among other things—a moment of trust. Whether or not you appear here, I am grateful to you. I'm grateful to Jen Smith and Rolando Huerta, who first offered me chances to set up the booth in the spaces they arranged, and to everyone who's invited me to set it up since then. I'm grateful to the Warren Neighbors' Network for allowing me to place what I've learned in service of the well-being of their town and its people.

My best friend Danielle Dreilinger has believed in me and my writing, vocally and ceaselessly, since she was sixteen and I was fourteen. During the writing of this book, she continued a thirty-year tradition of encouraging, reassuring, nudging, and celebrating me. She also connected me to my terrific agent, Jen Marshall, who championed this book and led me to my terrific editor, Renee Sedliar, who improved it tremendously.

Kate Colby and Darcie Dennigan midwifed this book with five years of alert, incisive feedback and unflagging encouragement, and with fifteen years of entwinement in each other's writing and lives. Nicole Hernandez Hammer supported the book not just as a contributor to the chapter where she appears but in talking through its concepts and its language with scrupulous, generous care.

Gratitude is due to Ann S., who nourished this book's beginnings and shared with me the crucial concept of *being organized by* others; to Mary-Kim Arnold, whose conversations with me compelled me to outline the project and whose conspiratorial humor helped me get through everything around it; to Ada Smailbegovic, whose communication with living creatures and understanding of "war mode" informed it deeply; to Lizz Malloy for wanting to hear and share the stories, and for principled rage; to Noraa Kaplan for modeling radical Jewish ritual and community as well as courageous, open-hearted writing; and to Guadalupe Elias and Stevie Redwood for online writing dates, productive challenges, pictures of soft animals, and truth-telling.

Nicole Soojung Chung, Maria Anderson, Jeff VanderMeer, Missy J. Kennedy, Haley E. D. Houseman, and Jennie Gropp all took editorial, collaborative, or supportive chances on the published writing of mine that preceded this book. On the other end of the process, Manu Shadow Velasco brought alert and patient attention to their sensitivity read of the almost-final draft.

Many people field-tested the questions and practices. Special gratitude for this time and attention is due to Isaac Sonnenfeldt and Stina Trollback; to Danielle Emerson; to Evan Donnachie, Hannah Fernandez, Ma'iingan Wolf Garvin, Carla Humphris, Vasu Jayanthi, Janek Schaller, Lily Seltz, Rose Shen, Isaac Slevin, and Caleb Stutman-Shaw; to Frances Vazquez; to Nathaniel Gan; to the Assembly of Light choir; and to Monster Trux. Gratitude of this kind is also due to Eva Amanda Agudelo, Aria Boutet, Rachel Hughes, and Janaya Kizzie for an early and especially illuminating field test, and for many years of friendship and ceremony. Correspondence and conversation with Londs Reuter, Hollis Mickey, and Lindsay Abromaitis-Smith helped make the practices more accessible to people with a range of abilities and disabilities; any failures in this area are mine.

Gratitude is due to The Department, among my first teachers in collectivity: Michael Tod Edgerton, Bronwen Tate, Caroline Whitbeck, and Lynn Xu. It's due to the organizations and groups I've learned with since, and their members: AMOR, No LNG in PVD, the Land and Water Sovereignty Campaign, and the Clary Sage gardeners. It's due to the original Interdependence Day crew, especially MacKenzie, Aria, Addie, and Emily. It's due to Indigo Bethea and Sovereign and Guinan Bethea-Gonzalez, to Remy Burnstein and Fiammetta Dimitri, to Benjamin Hammer, and to Owen Schapira. It's due to the lives and the memory of C. D. Wright, Mark Baumer, Olga Tabakman, Esther Ambrosino, and Juliette Goldenberg.

To my parents and sisters, Joel, Diane, Annie, and Rachel Schapira, and to my husband, James Kuo: you knew I could do it and you set me up to do it. My gratitude and love to you, every single day.

GLOSSARY OF WORDS AND PHRASES

Some of the words and phrases that can help us think, talk, feel, and act about climate change are uncommon; others are common, but have meanings that often shift around according to the speaker's beliefs and goals. Here is how I'm using them. Some of them I've also defined where they appear in the chapters.

Capitalism: The economic system that dominates our world right now, where living beings, objects, materials, and work are treated as sources of value to be extracted.

Climate change: Changes in the conditions of life on Earth because of the rising *average* temperature of the Earth's land, water, and air—meaning that more temperatures in more places are hotter than in the past, and creating drastic extremes and fluctuations in all kinds of weather. Sometimes also called global warming.

Climate collapse: A comprehensive breakdown of the systems that most people depend on for life. If it happens, it will bring widespread suffering and death.

Colonization: The practice of entering a place as a group, especially as members of a nation or company; exploiting what you find there, including soil, minerals, water, people, and other forms of life; and defending that exploitation with violence.

Ecofascism: Dictatorial and punishing control of people's lives and movements with "the environment" as an excuse. Often builds on and worsens existing hierarchies and systems of power (racism, ableism, etc.).

Ecosystem: A set of relationships among living beings and the conditions where they live.

Extractive: Describes practices, industries, and approaches that take without giving back, often leaving damage behind them.

FEMA: Federal Emergency Management Agency. In the United States, sometimes steps in with money, labor, or resources after a disaster.

Fossil fuels: Fuels that release heat-trapping gases (greenhouse gases) when humans burn them, leading to climate change, and that usually damage ecosystems when humans extract them from the ground. Much of the way that many people work and live in the present is also dependent on these fuels. Gasoline, oil, methane (natural gas), and coal are all fossil fuels.

Frontline: This word is used in two ways in this book. The first is the same as "frontline community," listed subsequently: a place where the effects of climate change are hitting or expected to hit first and worst. The second refers to the frontline of resistance against fossil fuel infrastructure (or other forms of large-scale damage) and the state violence that goes with it.

Frontline community: A place of people who get hit first and worst by the effects of climate change; this most often affects people who are poor, Black, Indigenous, and/or people of color.

Indigenous: Describes people who evolved with the land where they live or have lived, especially to distinguish them from people whose ancestors came to that land through conquest or colonization.

Infrastructure: Systems built and/or used by humans to move materials, people, work, and information. As an example, a river becomes part of infrastructure when we use it to transport large amounts of garbage, or when we ride on it to hunt, forage, or visit.

Just Transition: The transition from an extractive economy and way of living to one that nourishes, restores, and heals living beings, systems, and relationships. Includes generating less waste and pollution, making jobs and work that sustain and don't hurt the people who do them or the people who live near them, and learning how to live well in partnership with land, water, air, and other living beings. The Climate Justice Alliance introduced me to this term.

Land Back: Returning land to the Indigenous peoples whose ancestral lands it was or is. This can mean that the tribe or nation has the final say over what happens on the land and how humans relate to the rest of the land, that humans who live there pay their property taxes to the tribal government, or a range of other structures and relationships. Also sometimes called land return.

More-than-human being: A way to name animals, plants, fungi, bodies of water, and so on that distinguishes them from humans without suggesting that they're less alive or important. Zoe Todd's writing introduced me to this term.

Multisolving: An approach where people pool expertise, funding, and political will to solve multiple problems and create better conditions overall with a single investment of time and money. Elizabeth Sawin's work introduced me to this term.

Mutual aid: Sharing skills and meeting needs as equals, often among people who live near one another in order to build their power, and often to reduce dependency on harmful systems or institutions in order to weaken *their* power.

Redlining: Companies' practice of denying loans or insurance to people they think are a "poor financial risk"—usually because of racism, classism, and other factors that also make it difficult for people of color, disabled people, and generationally poor people to make a living. Often refers to home loans and insurance specifically, imposing segregation through money rather than law.

Sacrifice zone: An area, neighborhood, or region that industry and government have decided is acceptable to sacrifice to environmental damage—pollution, flooding, etc.—usually because they have contempt for the people who live there.

Somatic: Related to our bodies and the ways they're continuous with our minds. "Somatics" or "somatic experiencing" is a discipline and practice that guides people in processing histories and experiences through their bodies' sensations, and that practice has informed some of the exercises in this book.

State violence: When a government pays its employees to hurt people. Includes, but isn't limited to, military and police violence.

Stewardship: The practice of actively tending and nourishing the living systems we're part of, for the overall well-being of the beings in them.

Structure, structural: A structure is one name for a system that is built by people. It can be a structure of power, of value, of behavior or activity. When I describe something as structural in this book, I'm usually calling attention to the fact that it is widespread and is maintained by human activity.

System (political system, social system, living system . . .): A collection of elements connected in a way that produces a pattern of behavior or activity. Thanks to Elizabeth Sawin and Donella Meadows for this definition.

Tokenizing: When someone who's a member of an oppressed or marginalized group is treated by people around them as though they speak for that whole group, or as though their presence *in itself* indicates that group's involvement or approval in a decision or action, they are being tokenized. Think "the token Latina" or "the token disabled person" in a workplace, organization, or gathering of people who don't share those conditions.

Trauma: The aftereffects in our bodies and minds when something happens that injures our sense of self and/or our sense of the world. *Vicarious trauma* is when we are affected by the knowledge of suffering around us. *Collective trauma* is a weight on our spirits that people feel even when they don't consciously recognize the drastic change or loss that's causing it.

White supremacy: The attitude that white human lives matter more than any other kind of life, and the political, behavioral, and structural patterns and hierarchies that are informed by that attitude.

INDEX